IMAGES OF THE ICE AGE

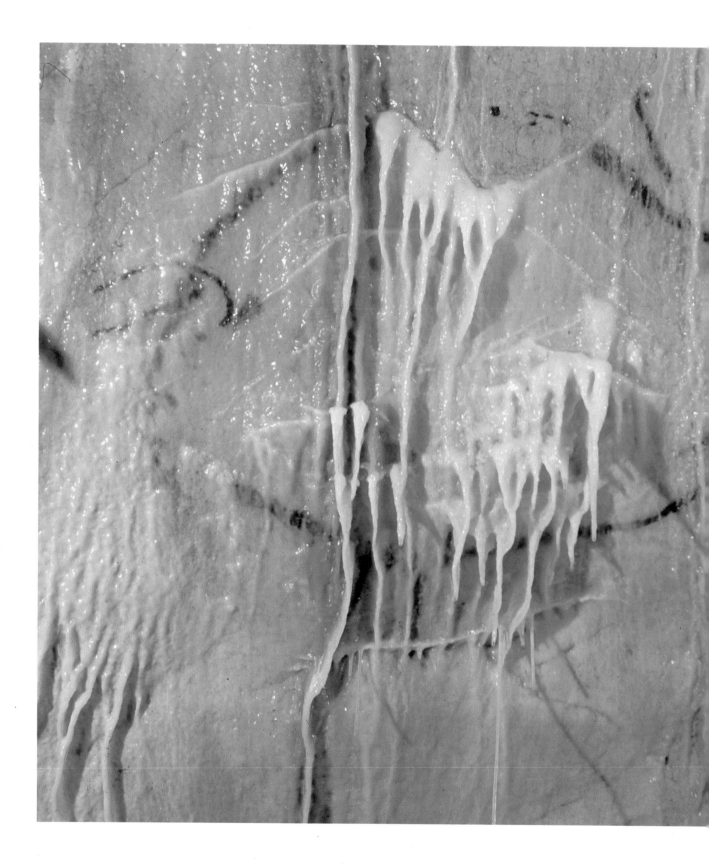

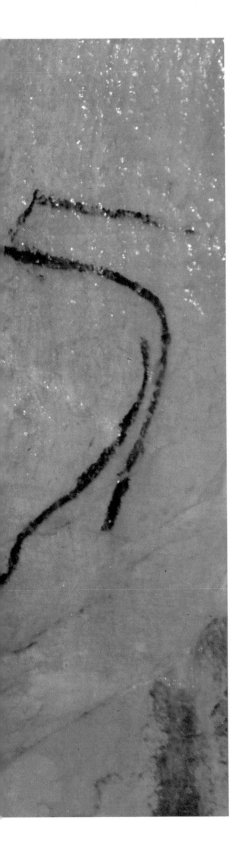

IMAGES
OF THE
ICE AGE

Paul G Bahn &
Jean Vertut

Facts On File
New York • Oxford

For Eric and Glyn
and
For Yvonne Vertut

Facts On File, Inc.
460 Park Avenue South
New York, New York 10016

Library of Congress Cataloging-in-Publication Data
Bahn, Paul G.
  Images of the Ice Age.

  Bibliography: p.
  Includes index.
  1. Paleolithic period – France   2. Paleolithic period –
Spain. 3. Cave-drawings – France.   4. Cave-drawings –
Spain   5. France – Antiquities.   6. Spain – Antiquities.
I. Vertut, Jean.   II. Title.
GN772.22.F7B32   1989   936.4   88-33405
ISBN 0-8160-2130-9

Facts On File books are available at special discounts when
purchased in bulk quantities for businesses, associations,
institutions, or sales promotion. Please contact the Special
Sales Department at 212 / 683-2244.
(Dial 1-800-322-8755, except in NY, AK, HI)

10 9 8 7 6 5 4 3 2 1

This book was designed and produced by
Bellew Publishing Company Limited
7 Southampton Place, London WC1A 2DR

Printed and bound in Italy by New Interlitho, Milan

Frontispiece:
Bison drawn in the Réseau Clastres (Ariège)
with concretions on top of it. Probably Magdalenian.
Length: 1.14 m. (JV)

Contents

Foreword

by Count Robert Bégouën

It was inevitable that the destinies of Paul Bahn and Jean Vertut would lead one day to their meeting on the path to a Palaeolithic sanctuary. Although they had totally different academic backgrounds, both from their youth had been attracted to and fascinated by the civilisations of the 'Reindeer Age'.

Paul Bahn rapidly dedicated himself totally to their study, travelling round Europe and, later, the world to obtain first-hand experience of archaeological sites, materials and prehistorians. How can one ever forget the sight of this young Englishman, thirsting for knowledge, arriving in Ariège on an overloaded old *vélomoteur*! From museums to libraries he took the time to see and read everything, a procedure that is quite exceptional in this age of specialists. The result was a doctoral thesis whose title sums up its approach: 'Pyrenean Prehistory: a palaeoeconomic survey of the French sites' (1984). In it he poses a thousand new questions about the various activities of prehistoric people, their movements, their way of life, their mentality, all of which are avenues of research. He does not hesitate to call into question some ideas which were considered solid, to highlight certain contradictions, compare the observations of different scholars, confront them with his own, and to produce a synthesis that is admirable for its finesse and its good sense. This enormous work has already become a precious tool for prehistorians.

After an education at the Ecole Centrale, from 1951 onward Jean Vertut took part in the excavations of Arcy-sur-Cure, directed by Professor André Leroi-Gourhan; his role was that of a specialist in colour photography underground. When the decorated zone of the Grotte du Cheval was discovered, Jean was naturally invited to take his first slides of parietal art.

Leroi-Gourhan very quickly noticed the originality of this talented and likeable young man who had an intellectual and spiritual curiosity that was always ready to marvel at *le Réel*. When the time came, he entrusted Vertut with the task of producing the illustrations for his master-work, *Préhistoire de l'Art Occidental*.

Fig. 2 *Jean Vertut photographing the 'weasel' (probably Magdalenian, length 46 cm) drawn in the Réseau Clastres (Ariège). He almost never used a tripod, he just stopped breathing to take the shots. (Photo R. Simonnet)*

Though he had an excellent 'eye', the principal quality of a photographer, Jean Vertut was also an engineer in telemanipulation and robotics; these disciplines also helped him to develop new, increasingly sophisticated procedures enabling one to approach and to photograph clearly the masterpieces of prehistoric art that were hidden away, sometimes in inaccessible crannies. He was a pioneer in the problems of focusing at a distance, computing light-measurements, and the use of simultaneous multiple electronic flashes connected to the camera – all techniques which have become run-of-the-mill today.

It was in 1956, during a visit to the cave of Trois-Frères, that the Bégouëns first made the acquaintance of Jean Vertut and of Yvonne, his wife, who was a precious assistant in his subterranean expeditions; but it was only in 1964 that Jean came to take his first series of slides in our caves. I was so enthusiastic about his results that I then asked him to undertake the exhaustive photographic coverage of the Volp Caves with me. This work, a long-term project, led us to spend hundreds of hours together, in conditions that were sometimes unbelievable but which enabled Jean's lens to capture the figures in their entirety, to eliminate as far as possible the deformations caused by the irregularity of the rock, and to bring out various little details that the artist had engraved carefully; all of this improved notably the reading and understanding of the drawings.

Ceaselessly perfecting his technique, Jean Vertut became *the* undisputed photographer of prehistoric art. Shortly before his death, he had started important research on 3-D photography as applied to prehistory, with possible extensions into photogrammetric recording. Thanks to him, photography of parietal art passed the stage of simple – albeit very beautiful – illustration, and became a separate technique of research.

Readers will take great pleasure in discovering that technique in this book. The magic of the images will give them wings, and everything will then be transformed, transcended! Everyone will feel something human and essential in these messages that were engraved, painted or sculptured in the caves during the twenty millennia of the Upper Palaeolithic. These images touch us, simply because they are beautiful. They remind us that people never cease to improve what they have undertaken to do. Whatever their motivations (and we know that the layouts are organised and complex), it is nevertheless clear that at some point people *loved* their works and, consciously or unconsciously, integrated Beauty into their dialogue with Matter.

Count Robert Bégouën
1988

Preface and Acknowledgements

Jean Vertut and I began planning this book in 1984. His sudden and untimely death in May 1985 at the age of only 56 robbed the world of a kind and remarkable man, and cave-art of its foremost photographer. It also meant that the book could no longer be what we envisaged, since Jean could neither choose the pictures himself, nor describe his methods and techniques as he had wished to do, nor photograph the new sites and objects that we had planned to include.

Nevertheless, with the help and encouragement of his widow, Yvonne, it has proved possible to produce an approximation of the original plan. Naturally, it has been necessary to draw primarily on the range of sites represented in Jean's existing collection, with only a few pictures borrowed from other sources where required to illustrate precise points. Inevitably, therefore, this book concentrates heavily on the Ice Age art of France and Spain.

It was never our intention, however, to produce an encyclopaedic work or a catalogue, or to attempt to replace the great tome by the late André Leroi-Gourhan (which remains one of the glories of Jean's career as well as of Leroi-Gourhan's); we wished to give a survey of past and current theories about the art rather than promote a single favourite interpretation. For reasons explained in the text, it will be some time before another comprehensive volume of that type can appear – if indeed such a thing is desirable. Sadly, our own book has now taken on an extra, unforeseen role, that of serving as a tribute to Jean by displaying the best of his final work. It goes without saying that, since the text was written in 1987, it is entirely my own responsibility; Jean may not necessarily have agreed with all the views expressed here, and the text is the poorer for the absence of his influence and advice.

Since the book is meant for the interested layman as well as for the academic, we decided not to get bogged down in arguments about the definition of 'art', or in speculations as to how and why it arose and disappeared. Much of that debate is redundant, since the earliest art to appear archaeologically is at an already well-developed stage, and we have lost its antecedents. In any case, these subjects have been treated elsewhere, in great depth.[1] Art, or at least the ability to create images in materials that have survived, appears relatively suddenly in the archaeological record at roughly the same time as *Homo sapiens sapiens*; clearly, something had developed in the mentality of our sub-species (and perhaps in that of its predecessors) which enabled or stimulated it to produce pictures, to adapt pre-existing shapes or to visualise new forms before they exist.

The term 'art' is rapidly losing favour among those researching the Palaeolithic period, since it presupposes an aesthetic function and lumps together a wide range of objects and types of image-making spanning 25 millennia; other vaguer terms – 'pictures', 'iconography', 'images', 'pictograms/ideograms', 'symbolic graphisms' or 'decoration' – are now preferred. In this book, therefore, the word 'art' is retained simply as a convenient blanket-term for the decoration of objects, rocks and cave-walls, whatever its function or motivation may have been.

Similarly, we chose not to devote the book to long, detailed and tedious discussions of the style and dating of particular caves, and thus to avoid all the entanglements and contradictions those subjective topics lead one into; at least one recent book has dwelt at length on these issues;[2] nor did we wish to devote much space to the equally impressionistic and subjective topic of what it feels like to visit these sites – this too has been done quite vividly in a recent popular work.[3] Moreover, it is superfluous to describe the beauty of particular depictions – readers can make their own judgements from the illustrations.

Very few books cover both the cave-art and the portable art of the last Ice Age, although the two cannot effectively be studied without reference to each other. We were determined to include both, but our book suffers from the usual archaeological vice of displaying only the finest examples; as such, they are not truly representative of the whole, but colour plates are a limited and valuable resource, and it seems a pity to waste them on illustrations which appear mediocre in execution, are in poor condition, or would be unintelligible to the untrained eye.

Lack of space has precluded setting the art within the cultural, economic and social context of Ice Age life, although such aspects inevitably crop up here and there. To integrate these themes properly would require a separate book; meanwhile, a number of texts about Ice Age life and culture are readily available to the layman, while the specialists will have their own views. Palaeolithic art includes so many subjects, and so many erroneous facts and interpretations are already in print about it, that it seemed best to cover it as fully and as factually as possible, in order to correct former errors and to assess the contribution made by recent work and discoveries. References to parallels in other cultures have been kept to an absolute minimum.

Since the text is aimed at the non-specialist, radiocarbon dates have been given as a straightforward figure – they should be taken as only a rough guide to the true age of the object or layer in question; specialist readers can find full details of the plus/minus range, the substance dated, and the laboratory concerned, by following up the references provided.

More and more people are becoming interested in the decorated caves, but fewer and fewer sites can be visited because of the sheer difficulties of access, or the potential damage to the images from accidents, vandals or bacteria – it is generally reckoned that a decorated cave open to the public is probably doomed. There is therefore a clear need for other ways of presenting the art to the public. Lascaux II, a copy almost indistinguishable from the original, is the ideal answer to this problem, albeit an extremely expensive one; for other sites, videos and television documentaries are a partial solution, and so are books of this kind. Much cave-art, however, can be fully appreciated only in three dimensions, so that one can see how the artist was inspired by or utilised the shape of the rock. Among Jean's last photographs are a series of stereoscopic studies which he hoped to place in this volume, together with special glasses to produce a 3-D effect. Such publications will inevitably appear within the next decade, thanks in large measure to the expertise and vision of this great pioneer.

My own acquaintance with Palaeolithic cave-art began in June 1963 when, at the age of nine, I heard the exciting story of the discovery of the art in the Volp caves (by the Bégouën brothers in 1912/1914) in a BBC radio broadcast for schools (*People, Places and Things*); at about the same time I acquired – as a free gift from some brand of washing powder – a waste-paper bin, which I use still, decorated with Lascaux-type animals! A more decisive influence was to come at Cambridge University where, during my first undergraduate supervisions with Eric Higgs, I was held spellbound by his tales of visits to the caves of France and Spain. Unlike some of his disciples, and in contrast with his own image as a prehistorian contemptuous of the dilettante art-historian approach to the past, Eric loved Palaeolithic art. My interest was heightened by reading books on the subject, and in lectures at Cambridge given by John Coles and Charles McBurney.

Finally, it was Eric Higgs who suggested that I do the research for a doctoral dissertation on the prehistory of the French Pyrenees – no doubt because he knew I would be fascinated by the region's richness in Palaeolithic art. During expeditions to southern France and Spain with him in the mid-1970s, I was agreeably surprised to find him making detours so that I could visit Gargas, Niaux, Le Mas d'Azil, Candamo, and other sites. After his death in 1976, I pursued my research in France with the active support and encouragement of the late Professor Glyn Daniel, who helped me to obtain my first permission to visit Lascaux, took a keen interest in my work and, I am proud to say, became a close friend. It is therefore to these two men, entirely different yet both admirers of Ice Age art, that I would like to dedicate this volume. I wish they had lived to see it.

During the year 1985/6 I was fortunate enough to be awarded one of the first J. Paul Getty Postdoctoral Fellowships in the History of Art and the Humanities, and used it to improve my knowledge of the cave-art sites and literature in Europe (using Toulouse as a base, through the kindness of Claude Barrière), as well as to travel abroad to see as much prehistoric art as I could, especially in Australia. I am profoundly grateful to the Getty Trust for such a wonderful opportunity, from which this book has benefited enormously.

In the course of my 'Getty travels' as well as during visits to France and Spain over the years, I have been helped by a great number of friends and colleagues in my research on Ice Age art. In Paris, the late Léon Pales, the late André Leroi-Gourhan, Arlette Leroi-Gourhan, Claude Couraud, Henri Delporte, Denis Vialou, Michel Garcia, Georges Sauvet, Dominique Buisson, Geneviève Pinçon, Lucette Mons, Suzanne de St Mathurin, Michel Orliac, Sophie de Beaune, Gilles Tosello, Jacques

Allain, Nicole Limondin, Marthe Chollot-Varagnac, Marina Rodna. In the Pyrenees, Robert Bégouën, Jean Clottes, Dominique Sacchi, Claude Barrière, Louis-René Nougier, Aleth Plenier, Carol Rivenq, Anne-Catherine Welté, Jacques Omnès, Jean Vézian, Jacques Blot, Robert Arambourou, André Clot, René Gailli, Luc Wahl, Romain Robert, Jean Abélanet, Robert Simonnet, Georges Laplace, Roger Séronie-Vivien, Claude Andrieux. In the Périgord, Brigitte and Gilles Delluc, Paulette Daubisse, Jean Gaussen, Alain Roussot, Jean-Philippe Rigaud, Jean-Marc Bouvier, Jean-Michel Mormone. In Spain, Jesús Altuna, Ignacio Barandiarán, Vicente Baldellou, Reynaldo González García, Alfonso Moure Romanillo, Francisco Jordá Cerdá, Josep Maria Fullola Pericot, Jose Manuel Gómez-Tabanera, Joaquín González Echegaray, Eduardo Ripoll Perelló, José Miguel de Barandiarán. Elsewhere in Europe, Lya Dams, Marylise Lejeune, Yves Martin, André Thévenin, Martine Faure, Jean Combier, Michel Lorblanchet, Gerhard and Hannelore Bosinski, Joachim Hahn, Karl Narr, Hans Biedermann, Bohuslav Klíma, Jan Jelínek, Zoïa Abramova, Mike and Anne Eastham, Pat Winker, Warwick Bray, Alex Hooper, Jonathan Speed; and outside Europe, David Lewis-Williams and Anna Belfer-Cohen. In America, Alex Marshack, Meg Conkey, Ray and Jean Auel, Tom and Alice Kehoe, Glen Cole, John Pfeiffer, Noel Smith, Whitney Davis. In Australia, Robert and Elfriede Bednarik, Geoff Aslin, John Clegg, Pat Vinnicombe, Sylvia Hallam, Andrée Rosenfeld, George Chaloupka, Jo Flood, Rhys Jones, Percy/Steve/Matt Trezise, Paul Taçon, Jo McDonald, Lesley Maynard, Charles Dortch. In China, Zhu Qing Sheng, You Yuzhu; and in Japan, Yuriko Fukasawa, Yoshiaki Kanayama, Gina Barnes and Hideji Harunari.

For supplying special pictures for this book I am particularly grateful to Dominique Sacchi, Robert Bégouën, Robert Simonnet, Carol Rivenq, Claude Couraud, Gerhard and Hannelore Bosinski, Alfonso Moure Romanillo, Glen Cole, Andrée Rosenfeld, Brigitte and Gilles Delluc, Jean Clottes and Ulm Museum.

Special thanks go to Meg Conkey, Whitney Davis and John Pfeiffer for their support and for detailed comments on the typescript.

The original idea for a volume co-authored with Jean Vertut came from Ib Bellew; the project was initiated with the help of Brigitte and Gilles Delluc, set up by my agent, Andrew Best, and brought to fruition thanks to the unfailing help of Yvonne Vertut. Deirdre McDonald showed remarkable skill in transforming text and pictures into a coherent book. I am deeply grateful to my – and Jean's – good friends, Robert Bégouën and Alex Marshack, for writing the Foreword and Appendix.

Finally, I have a tremendous debt of gratitude to the Getty Grant Program of the J. Paul Getty Trust for following up my postdoctoral Fellowship with a publication grant towards the production of this book, which therefore owes its final form and, to a considerable extent, its very existence to the Trust's generosity and support.

PAUL G. BAHN
1988

Introduction: Cave Life

Despite persisting popular belief, fuelled by the best efforts of movie-makers and cartoonists, it is not true that people during the late Ice Age habitually lived in caves. Over the last hundred years, archaeological investigation has made it clear that they primarily occupied cave-mouths, sunny rock-shelters, and huts or tents in the open air. This is hardly surprising, since caves are not the most pleasant choice of habitat – they are usually dark, wet, slippery, and full of rocks and jagged concretions.

Nevertheless there are cases, many of them in the French Pyrenees, where hearths and other signs of occupation have been found far inside very deep caverns, and even hundreds of metres from the entrance (assuming that the present entrance was also the main one in prehistory). What is the explanation for this? In some cases these are temporary encampments linked to the creation of decoration on the cave-walls: for example, in the cave of Tito Bustillo in northern Spain, occupation debris and colouring materials have been found in excavations at the foot of the principal painted frieze (p.97).

In other cases they may denote the seeking of refuge in particularly harsh conditions: at present the caves of southern France and northern Spain maintain a fairly constant temperature (usually c 14°C) which makes them pleasantly cool in summer and mild in winter. However, recent work by geophysicists has established that during the last Ice Age it was not merely mild but quite warm inside the caves – a fact which helps to explain why the Palaeolithic footprints found in their depths are of bare feet and have no sign of frostbite[1] (but see p.105 for the possibility of frostbite on hands).

The footprints are quite often those of children: for example in the cave of Aldène (Hérault), or in Fontanet (Ariège) where a child – who also left knee- and handprints – seems to have pursued a puppy or a fox into the cave's depths. Children were clearly not afraid to explore the far depths, narrow passages and tiny chambers of caverns, whether alone (as at Aldène) or with adults; it is possible that the finger-holes and heel-marks found around the clay bison of the Tuc d'Audoubert (Ariège) were made by children playing while the adult(s) made the figures.

Fig. 3 *Prehistoric footprints in Niaux (Ariège), carefully and deliberately made by two young children. Precise date unknown. (JV)*

It is a tragedy that we have irretrievably lost huge amounts of fascinating information from cave-floors: in some cases, the caverns were frequented throughout history, and thus any prehistoric traces on the floors were destroyed centuries ago; in others, the prehistoric galleries were discovered during our own century, but the discoverers inadvertently obliterated the precious evidence because (no doubt with their attention focused on the decorated cave-walls) they simply did not notice the footprints, objects and even engravings at their feet. With one important exception – the Tuc d'Audoubert – it is only in the most recently discovered caves such as Fontanet and Erberua that all such traces have been carefully preserved intact.

The Tuc d'Audoubert's inner depths were first explored in 1912; as the cave is privately owned by the Bégouën family, visits have been kept to a strict minimum ever since, and nothing has been disturbed. Thus the cave not only has artistic treasures (above all, the unique clay bison figures, as well as engravings), it is also a treasure-house for all manner of evidence about the activities of the Palaeolithic visitors: flint tools, teeth or pieces of bone were carefully placed in rock-crevices, or stuck into the floor; stalagmites were deliberately broken; cave-bear jaws were picked up, their canines were removed (presumably for use in necklaces) and then the jaws were thrown down again; the traces in clay around the bison have already been mentioned. In short, since no one visited these galleries between the last Ice Age and 1912, one can actually follow the traces left by Magdalenian people and reconstruct many of their actions (see Appendix).

The same applies at Fontanet, a gallery blocked during the Magdalenian period and only rediscovered in 1972. Here, the back part of the cave has no traces of occupation but does have numerous prints, as mentioned above. The front part near the blocked entrance has no prints, but it has engravings and paintings on the walls, and, on the floor, a series of hearths around which are the animal bones that the occupants probably tossed over their shoulders. Here again, the vestiges and prints are so fresh that

Fig. 4 *Prehistoric footprints of three children in the Réseau Clastres (Ariège). (JV)*

one would think they were made a few minutes ago rather than 13,000 or 14,000 years ago.[2]

Apart from boulders which may have been used as seats, no cave 'furniture' has survived. Presumably it was all made of wood and has therefore disintegrated through time – as yet, no one has been fortunate enough to find a waterlogged Palaeolithic site with preserved wood: only a few Palaeolithic wooden objects, including a couple of spears, have survived in Europe and elsewhere.[3] But we know that Palaeolithic people were perfectly capable of working wood – as will be seen in a later chapter, some of the decorated cave-walls definitely required ladders or scaffolding, and the actual sockets for scaffolding-beams survive in Lascaux.

In addition, other forms of evidence show us some improvements and amenities which were brought to the caves: at Enlène (Ariège), for example, thousands of small sandstone and limestone slabs were brought in from local sources and laid down as a kind of pavement; and, as we shall see, many hundreds of them were engraved. Pollen analysis of sediments in caves such as Lascaux and Fontanet has revealed that great clumps of grasses and summer flowers were brought in, presumably for bedding and seating.[4] So even though few caves were actually inhabited by late Ice Age people for any length of time, it seems they knew well how to adapt this environment to their advantage and comfort.

Life in the Upper Palaeolithic could be harsh or pleasant, depending on climatic and environmental fluctuations. At times there was plenty of food even in the areas close to glaciers, since periglacial environments are rich in flora and fauna, and there is no evidence of rickets in skeletons of the period. On the other hand, some skeletons display evidence of phases of slow growth caused by starvation or illness. Only a few cases of illness have been detected, such as hydrocephaly in a child (Rochereil, Dordogne), or a fungal infection in the old man of Cro-Magnon. Whereas the average Neanderthaler died aged about 30, and very few of them passed their mid-40s, Upper Palaeolithic people had an average age at death of 32, and some were well over 50.[5]

This may not seem a great age, but death between 30 and 40 was normal until only a few centuries ago, and indeed remains so in some parts of the world. On the whole, Upper Palaeolithic people were fit and healthy; little evidence has been found for arm or leg fractures, and indeed the only indication of any violence in this period in Europe (apart from the alleged 'pierced men' depicted in two caves; see p.152) is a woman from Cro-Magnon with a deep cut at the front of her skull (a wound to the brain which she nevertheless survived for some time); seven skulls from the upper cave of Zhoukoudian in China have enigmatic depressed fractures.[6] On the whole, therefore, life was peaceful in the Upper Palaeolithic, as far as we can tell.

Since this book is primarily about the art of Franco-Cantabria, a brief guide to the cultures and their dates is required. The Middle Palaeolithic (period of Neanderthal people), also called Mousterian after the site of Le Moustier (Dordogne), was succeeded by the Upper Palaeolithic (period of Cro-Magnon people) which has been divided into a series of phases named after different sites in France. The following is a very rough guide to their dates:

Châtelperronian:	starts *c* 35,000 BC
Aurignacian:	*c* 30,000 BC
Gravettian:	*c* 25,000 BC
Solutrean:	*c* 20,000 BC
Early Magdalenian:	*c* 15,000 BC
Middle Magdalenian:	*c* 13,500 BC
Late Magdalenian:	*c* 10,000 BC
Azilian:	*c* 8,000 BC

The Magdalenian has also been divided into a sequence of phases (I to VI) which have to be mentioned occasionally in the text, although they are rapidly being abandoned by prehistorians.

1

The Discovery of Ice Age Art

The saga of how we discovered and came to terms with the fact that our remote ancestors could be brilliant artists is a fascinating one, filled with missed chances, pioneering insights and, above all, the interplay of open and closed minds which has always dogged archaeology, a subject in which schools of thought and personal antagonisms, even today, can play as great a role as the actual evidence.

Portable objects

The first pieces of art from the Palaeolithic (Old Stone Age) were found in about 1833, in the Magdalenian cave of Veyrier (Haute-Savoie), near France's border with Switzerland, where a certain Dr François Mayor of Geneva encountered a pseudo-harpoon of antler, engraved so as to resemble a budding plant, and a perforated antler baton decorated with a simple engraving which may perhaps be a bird (Fig.5);[1] another more famous baton from the same site, with an engraving of a plant on one side and an ibex on the other, was found only in 1868. A horse-head, engraved on a reindeer antler, is said to have been found in 1842 at Neschers (Puy-de-Dôme);[2] and a reindeer foot-bone, bearing a fine engraving of two hinds (Fig.6), was discovered by Joly Leterme (an architect) and André Brouillet (a notary) in the French cave of Chaffaud (Vienne) in 1852.[3]

At this time, however, the study of prehistory was in its infancy, and there was as yet no concept of an Old Stone Age, let alone of Palaeolithic art; thus the Chaffaud engraving, for example, was declared to be Celtic in date. It was not until some years later that, through comparison with examples excavated in Palaeolithic occupation layers elsewhere, the Chaffaud bone was properly identified.

The existence of Palaeolithic art was first established and accepted through the discovery, in the early 1860s, of engraved and carved bones and stones in a number of caves and rock-shelters in South-West France, particularly by Edouard Lartet, a brilliant French scholar funded by Henry Christy, a London banker and industrialist. After finding his first piece of portable art in 1860 (a bear's head engraved on antler, from the

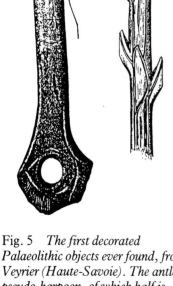

Fig. 5 *The first decorated Palaeolithic objects ever found, from Veyrier (Haute-Savoie). The antler pseudo-harpoon, of which half is shown here, is 11.2 cm long. (After Pittard)*

cave of Massat in the French Pyrenees) Lartet published a drawing of it in 1861 together with a sketch of the Chaffaud bone; he was thus the first person publicly to identify the latter as Palaeolithic.[4]

Fig. 6   Detail of the reindeer foot-bone from Chaffaud (Vienne), showing engraved hinds. Probably Magdalenian. Total length: 13.5 cm, height 3.7 cm. (JV)

The depictions he encountered in his excavations came as a great surprise: their quality was astounding, since it had been assumed that prehistoric people were primitive savages with no leisure time and no aesthetic sense. Yet there could be no doubt that the objects were authentically ancient – they were, after all, associated with Palaeolithic stone and bone tools, and the bones of Ice Age animals; some of them, such as the magnificent mammoth engraved on a fragment of mammoth ivory from La Madeleine,[5] accurately depicted extinct species while others, such as the reindeer from Bruniquel (Fig.17), showed animals which had long since deserted this part of the world.

The phenomenon was so new and unexpected that museums took some time to make up their minds about it. For example, when the Vicomte de Lastic-Saint Jal offered to the Louvre the art objects which he had dug up at Bruniquel in 1863, the museum hesitated – in part, because of his high price – and the British Museum snapped them up, subsequently buying others from this site and elsewhere. By 1867, however, Palaeolithic art was sufficiently well established for 51 pieces to be included in the 'History of Work' section of the Universal Exhibition, in Paris.

These first discoveries triggered a kind of 'gold rush' which saw numerous people excavating – or, more usually, plundering – likely caves and shelters in search of ancient art treasures. Little attention was paid, except in a few cases, to the exact stratigraphic position of these finds; they were dug up like potatoes. At the same time, however, some diggers were ignoring the art on the cave-walls. In some cases they must have seen it, but they did not *observe* it: in Archaeology, one tends to find what one expects to find, see what one expects to see. In other words, the cave-art was of no importance in their eyes because it could not be ancient: it was inconceivable that it might have survived so long, and nothing of the sort had yet been found. Nevertheless, a few pioneers had noticed cave-art and wondered...

Cave-art

Many caves in Europe, as elsewhere, have been frequented since the Ice Age and throughout history by new occupants, by shepherds, by the curious or the adventurous. In some areas, such as the Basque country, superstitions and religious traditions associated with caves are probably extremely ancient, perhaps even extending back to the period of Palaeolithic art. Certainly there is some evidence that rituals associated with the art may have survived in some areas until quite recently: in 1458, Pope Calixtus III (a Borgia from Valencia) prohibited religious ceremonies in 'the cave with the horse pictures'. We do not know which cave he meant, but from the description it is likely that its pictures dated to the Ice Age.[6]

Many decorated caves also contain graffiti of different periods – Gargas, for instance, has engraved marks on its walls which are probably protohistoric, Gallo-Roman and medieval in date. In some cases, it is certain that people saw the art: in the Pyrenean cave of Niaux, a certain Ruben de la Vialle wrote his name in 1660 on the only panel of the 'Salon Noir' left undecorated by the Magdalenians; later graffiti there are placed within or between animals, carefully avoiding the painted lines.[7]

Seeing the art is one thing; reporting it is quite another, and no cave-art, as far as we know, was ever mentioned in print before the nineteenth century.[8] This is quite understandable, since prehistory did not 'exist' until then, and so the pictures had no significance. However, once prehistoric studies got under way and portable art of the Ice Age had been discovered and authenticated, a few scholars at last began to notice what had been staring them in the face for so long.

The first definite suspicions of this type can be attributed to Félix Garrigou, a scholar who investigated scores of caves in the French Pyrenees in the 1860s and who found some pieces of portable art, including the famous bear engraved on a pebble, from the cave of Massat. During a visit to Niaux in 1864 he was taken to the Salon Noir, destined to be recognised as one of the glories of Ice Age cave-art (Fig.119). As

Fig. 7 *Photomontage of the painted ceiling at Altamira (Santander), covering roughly 18 m by 9 m. The hind on the left is 2.25 m long, while the curled-up bison are 1.5 to 1.8 m in length. Magdalenian. (JV)*

mentioned above, the Salon had often been visited in the past, and the local guides even called this part of the cave the 'museum'! In his notebook Garrigou wrote that 'there are paintings on the wall. What can they be?'[9] Alas, he did not make the crucial mental leap, and Niaux therefore narrowly missed being the first true discovery of Ice Age parietal (wall) art (its figures were not rediscovered until 1906, although a local schoolteacher called Claustre had mentioned the 'Salle des Bêtes' of Niaux in a note in 1885).

A few years later, in 1870, the archaeologist Jules Ollier de Marichard wrote in his excavation notebook that he had seen 'signs of the zodiac' engraved in a deep gallery of the cave of Ebbou (Ardèche), and in 1873 he added that he had seen animal silhouettes sketched on the walls.[10] But, like Niaux, Ebbou had to wait a considerable time (until 1946!) before being rediscovered. In 1879 he wrote a letter to Emile Cartailhac stating that some caves in Ardèche had red paintings of 'fantastic animals',[11] but unfortunately he put nothing in print.

One scholar, however, did print something about wall-art: in 1878, Léopold Chiron, a schoolteacher, noticed deep engravings in the cave of Chabot (Gard); he not only took photographs and imprints of them, but also published a note about them, although he could not know their date.[12] He mistakenly thought he could see birds and people among the lines; unfortunately, the Chabot engravings are difficult to decipher, and the figures are far from clear.

However, a more decisive event occurred in the same year, for it was at the Paris Universal Exhibition of 1878 that Don Marcelino Sanz de Sautuola saw Palaeolithic portable art on exhibit, an experience which led eventually to his discoveries at Altamira. Unfortunately, whereas the pictures in Chabot were too modest to make an impact, those of Altamira were too splendid to be believed.

Altamira

The cave of Altamira, near the north coast of Spain, was found in 1868 by a hunter, Modesto Cubillas, on freeing his dog which had become wedged in some rocks. De Sautuola, a local gentleman and landowner, first visited it in 1876; he noticed some black painted signs on a wall, but thought little of them. His subsequent trip to the Paris exhibition left him profoundly impressed and excited by the implements and art objects of the Ice Age on exhibit there – we know that he was drawn back again and again to those displays. He was also fortunate enough to meet Edouard Piette, the greatest prehistorian in France, who was in the process of amassing a remarkable and unrivalled collection of Palaeolithic portable art; the two men discussed caves, artefacts and excavation methods.[13]

Back in Spain, de Sautuola began his own excavations in caves, and eventually returned to Altamira in 1879, when he was 48 years old. As is well known, during November that year he was digging in the cave-floor, searching for prehistoric tools and portable art of the sort he had seen in Paris, while his little daughter Maria (according to her own account) was 'running about in the cavern and playing about here and there ... Suddenly I made out forms and figures on the roof'.[14] She exclaimed '*Mira, Papa, bueyes!* [Look, Papa, oxen!]'.[15] She had seen the cluster of great polychrome bison on the ceiling (Fig.7), which had lain concealed in total darkness for 15,000 years, awaiting her.

Her father was, to say the least, astonished. At first he laughed, then he became more interested, particularly when he found that the figures were done with what seemed to be a fatty paste. Finally 'he was so enthusiastic that he could hardly speak'.[16] He must have noticed at once the close similarity in style between these huge figures and the small portable depictions with which he was familiar, but we do not know exactly when he dared make the deduction that this wall-art was of equal age. He certainly knew that it had not been done since the cave's discovery in 1868, because no stranger could spend so much time and effort in there (and to what purpose?) without everyone in the area knowing about it.

In any case, de Sautuola's greatness lies in the fact that he not only made the crucial deduction, he dared to present his views to the sceptical academic establishment. First, he wrote to Professor Juan Vilanova y Piera, Spain's foremost palaeontologist, in Madrid. Vilanova visited him, dug a little in the cave, and was convinced that de Sautuola was right about the paintings. As a result of a lecture given about Altamira by Vilanova in Santander, it made front-page news all over Spain, and tremendous crowds of people came to see the cave. Even King Alfonso XII visited Altamira, crawling into the smallest galleries, which were lit by the candles of his entourage of servants, and leaving his name in candle-black on the wall near the entrance!

In 1880 de Sautuola published a booklet on his discoveries at Altamira, including the Ice Age paintings;[17] it was a cautious piece of work in which he did not affirm the contemporaneity of the painted ceiling and the Palaeolithic deposits, but merely posed the question. In the same year, Vilanova presented the discovery at the International Congress of Anthropology and Prehistoric Archaeology in Lisbon, a gathering of some of Europe's greatest prehistorians – among them Montelius, Pigorini, Virchow, Lubbock, Evans and Cartailhac.

One would think that such a remarkable find, albeit unprecedented and unexpected, would have met with interest – even excitement – at this gathering; that there would have been a stampede of specialists to see the site; that de Sautuola would have been fêted and congratulated. It is to archaeology's undying shame that the very opposite occurred, and that it helped to bring about his premature death in 1888, a sad and disillusioned man still under suspicion of fraud or naivety, with his discovery rejected by most prehistorians. He had taken the hostility personally, as an attack on his honour and his honesty.[18]

There were a number of reasons for the rejection; for a start, he was a complete unknown, rather than an established prehistorian. Secondly, nothing remotely similar had been found before; after all, figures scratched on bone were a very different phenomenon from sophisticated polychrome paintings. Moreover, the discovery was made in Spain – at this time, many of the leading prehistorians were French, and South-West France was the region *par excellence* for Palaeolithic finds. So it must have seemed illogical and suspect for a discovery such as Altamira to occur in a different area.

At the Congress, Vilanova exhibited drawings of the Altamira figures in the halls; but his presentation was met with incredulity and even an abrupt and contemptuous dismissal of the very idea – Cartailhac, one of the most influential French scholars, had been warned by his virulently anti-clerical friend, Gabriel de Mortillet, that some anti-evolutionist Spanish Jesuits were going to try to make prehistorians look silly. Where Altamira was

concerned, therefore, Cartailhac smelt a rat, rather than a bison, and walked out of the conference hall in disgust. When writing to de Sautuola in 1880 to acknowledge receipt of his booklet, he even informed him that his paintings of 'aurochs' (wild cattle!) were unlike the prehistoric animals, and had differently shaped horns![19]

It was claimed that the paintings were far too good to be so ancient, and that a modern artist (either duping de Sautuola or conspiring with him) had faked them. Cartailhac referred to the affair as a '*vulgaire farce de rapin*' (a dauber's vulgar joke). Yet none of the objectors had seen the original pictures. Vilanova issued formal invitations to the leading

Fig. 8   *Photomontage of the painted panel at Marsoulas (Haute Garonne), showing bison, horses and barbed signs. Probably Magdalenian. Total length: c 5 m. Note the bison composed of red dots (length: 87 cm). (JV)*

prehistorians to visit the cave – but, incredibly, all refused.

In 1881, one Frenchman, the engineer Edouard Harlé, did turn up to examine the art of Altamira; but, as an emissary of Cartailhac and de Mortillet, he had clearly made up his mind beforehand that the paintings had been done in the 1870s, between de Sautuola's two visits. His judgement, published in the most influential journal of prehistory of the time,[20] was based on a number of factors: the pictures were too good to have been done by prehistoric savages (this despite the quality of the authenticated portable art!); they were anatomically inaccurate (here he expanded Cartailhac's views about the anatomy of aurochs, totally irrelevant since the Altamira animals are clearly bison!); the paint looked too fresh to be ancient, it came away on the finger [sic], some of it had been applied with a modern paintbrush, and the cave was too humid and the rock too friable to have preserved art for so long; some of the figures were on top of stalagmite, while others were covered by only a thin layer of it, which need not denote great age (this is perfectly true); in addition, prehistoric artists would have spent long periods of time here, far from daylight, and the soot from their lamps would have blackened the ceiling, which was not the case. This last objection was more reasonable, given our knowledge at the time (no Palaeolithic lamps had yet been authenticated, although a few had already been found[21]); but, as will be seen (p.109), experiments now suggest that animal fat in lamps gives off no soot.

Cartailhac was pleased to publish Harlé's account, since he was irritated at the persistence of the two Spaniards at different meetings;[22] his own report of the Lisbon meeting had a few disdainful words about Altamira, while the official records of the Congress did not mention it at all! Other anti-Altamira articles began to appear in scientific journals and in the newspapers, both in France and in Spain. De Sautuola tried to present his case again at a French Congress in Algiers in 1882, but in vain; he submitted his booklet and a report to an international Congress in Berlin the same year, but no discussion was held. Finally, Vilanova gave up the fight, and de Sautuola stood alone.[23] Neither de Mortillet's book *Le Préhistorique* (1883), nor Cartailhac's *Les Ages Préhistoriques de l'Espagne et du Portugal* (1886), nor the International Congress in Paris in 1889 mentioned Altamira at all.

However, it would be wrong to give the impression that all the leading French prehistorians shared the opinions of Gabriel de Mortillet and Emile Cartailhac, the two most virulent and powerful voices against Altamira. For example, Henri Martin (grandfather of the famous prehistorian of the same name) had to decline an invitation to visit the cave in 1880, but said in his letter to Vilanova that he saw definite analogies between the drawings and portable art;[24] and Edouard Piette, who had met de Sautuola in 1878, was in no doubt about Altamira, since he had always thought that portable art must have been accompanied by wall-art[25] – indeed he wrote to Cartailhac in 1887 stating that he thought the paintings were Magdalenian; and in the late 1880s, when he discovered the painted pebbles of the Mas d'Azil which characterised the 'Azilian culture' at the very end of the Ice Age, he presented them as the oldest known paintings in France, matched or surpassed in age only by Altamira.[26] Needless to say, this discovery too produced howls of rage from the establishment until its validity was proved. Piette was courageous, imaginative and open-minded throughout his career, and stands among the true pioneers of early French prehistory.

Acceptance at last

How, then, did cave-art change from a source of ridicule to a subject worthy of splendid monographs? It is no coincidence that the metamorphosis took place in South-West France. After the Altamira débâcle, all was quiet on this academic front, but discoveries were still being made: the stratigraphic position of Piette's pebbles, for example, proved that ochre could adhere to rock for millennia.

The real breakthrough – literally – came at the cave of La Mouthe (Dordogne), where in 1895 the owner decided to remove some of the fill and exposed an unknown gallery; a group of four youngsters led by Gaston Berthoumeyrou explored it on 8 April, returning on the 11th with Gaston's father, Edouard; in the course of one of these visits, they spotted a bison engraved on one wall, 100 metres inside the cave, as well as other figures; because of Palaeolithic deposits in the blocking fill, it was clear that the pictures must be ancient.[27]

Emile Rivière – who had just visited Altamira – came to the cave and carried out some excavations, in the course of which he found more parietal pictures, and in 1899 unearthed a Palaeolithic lamp here, the first to be accepted,[28] and still among the finest because of its carved red sandstone and its engraving of an ibex. So La Mouthe provided proof of age, and of a lighting system. Cave-art could no longer be denied.

Meanwhile, Chiron had again called attention to his discovery of engravings in Chabot in 1889 and 1893; while in 1890 he also mentioned engravings in the nearby Grotte du Figuier (Ardèche). Moreover, in 1881, a gentleman named François Daleau had begun digging in the cave of Pair-non-Pair (Heads-or-Tails) near Bordeaux and found bones of Ice Age species; in 1883 he first noticed engravings on the cave-walls exposed by the removal of occupation layers, but he paid little attention to them, apart from mentioning the fact in his notebook. It was not until 13 years later that the discoveries at La Mouthe led him to clean the walls with a water-spray from the vineyards and study the pictures – he drew sketches of the animal figures in his notebook, and published an article about them.[29] Since some of them had been covered by Gravettian occupation layers, it was indisputable that the pictures were at least as old as that Upper Palaeolithic culture.

Daleau was visited by Piette, and also by Cartailhac who admired the engravings and tried to photograph them. Cartailhac was growing troubled: during a visit to La Mouthe, he himself had removed some Palaeolithic sediment and exposed the legs of a painted animal on the wall.

Further crucial revelations awaited him in the Pyrenees. A clergyman, the abbé David Cau-Durban, had been digging in the Pyrenean cave of Marsoulas since 1883, where he had found cultural material and portable art of the Upper Palaeolithic. The cave is small and narrow, and one wall is festooned with engravings and red and black paintings. The paintings, at least, are very visible; Cau-Durban had seen them but assumed they were modern, like the names and dates on the wall, even though some were partly covered by sediment.

After the discoveries at La Mouthe and Pair-non-Pair, Félix Regnault, a local scholar, returned to Marsoulas in 1897 and discovered the animal pictures. But his claims were greeted with hilarity: Cau-Durban mocked him, saying that he was naive and gullible, and that the pictures had been done by children;[30] although Regnault mentioned the paintings at a session of the *Société Archéologique du Midi*, stressing the analogy between the portable and the wall-art in the cave, Cartailhac, cave-art's old adversary, refused to publish a note about it in the Society's journal! Regnault got two lines, with a possible misprint that described the paintings as historic rather than prehistoric. Invitations were issued in 1898 to see the paintings, but Cartailhac did not come. Rivière, however, did pay a visit, and found the colours rather fresh – consequently he had doubts about the age of the paintings, though he accepted the engravings. He may have been influenced by Cau-Durban and Cartailhac, both of whom told him they had seen no animal paintings during the excavations. Ironically, as we shall see, it was a visit to Marsoulas in 1902 which finally opened Cartailhac's eyes.

In 1901, one of Rivière's diggers, Pomarel, had found engravings in the cave of Les Combarelles, near Les Eyzies (Dordogne), and told his teacher, Denis Peyrony, about them. The latter visited the cave with Louis Capitan and the young Henri Breuil, and found many more pictures. A few days later, Peyrony found the art in the nearby cave of Font de Gaume. These discoveries, together with all that had gone before, finally penetrated Cartailhac's stubbornness. In 1902 he visited Marsoulas and from there he went to Altamira with Breuil and at last saw its art for himself – after which he became one of the most enthusiastic scholars of cave-art, even purchasing Marsoulas in order to ensure its protection!

As mentioned earlier, Piette had written to him about Altamira in 1887; once cave-art began to be accepted at places such as La Mouthe, he had again reminded Cartailhac about Altamira, comparing it with La Mouthe. Yet in later years, Cartailhac claimed that he himself had re-opened the debate; he boasted that he discovered many engravings at Marsoulas which Regnault had missed;[31] above all, he published his famous article of 1902, '"Mea culpa" d'un sceptique',[32] which is still cited in France as the epitome of an objective scientist admitting his mistakes. Certainly its title and its intentions are laudable. However, on close inspection, the paper contains only limited contrition: indeed, Cartailhac states that his actions in 1880 were fully justified at the time, and that there is nothing to object to in Harlé's 1881 report. It is clear that Cartailhac changed his mind at the last minute only when the evidence became overwhelming – it was thanks to the new discoveries that 'we no longer have any reason to doubt the antiquity of Altamira'.

Harlé revisited Altamira in 1903, having seen Font de Gaume, and realized his mistake, though somewhat grudgingly; he wrote to Cartailhac: 'it is the Font de Gaume that has made me change my mind. But for that I would still say the Bison Ceiling is a forgery'.[33]

Cartailhac later admitted that he should have expected parietal art to exist, once portable art had been discovered, and he blamed early opposition to Altamira on the fact that 'we were blinded by some dangerous spirit of dogmatism';[34] his 'Mea Culpa' of 1902 was reported by the whole Spanish press with satisfaction.[35] De Sautuola was vindicated.

On 14 August 1902, a number of prehistorians – including Emile Cartailhac – attending the Montauban Congress of the French Association for the Advancement of Sciences made an excursion to the Les Eyzies area and visited the decorated caves there. This marked the official recognition by science of the existence of cave-art. Had more minds been open and willing to accept the unexpected, had Chabot been more splendid and Altamira less so, cave-art studies could have begun two decades earlier. However, they quickly made up for lost time.

2

A Worldwide Phenomenon

After these first discoveries, examples of portable art and new decorated caves continued to be found in Europe (as will be seen, below); but for a long time it was believed that, apart from the material in Siberia, Ice Age art did not exist elsewhere, and was an exclusively European phenomenon. Little by little, however, and especially in the last few years, it has become apparent that, towards the end of the Pleistocene, artistic activity was underway all over the world. The technical, naturalistic and aesthetic qualities of European Palaeolithic images remain almost unique for the moment, but it is still true that, at this period, in other parts of the world, one can see traces of the same phenomenon.[1]

The New World

Amazingly, the first clue to Pleistocene art in America was found in 1870, only a few years after Lartet's finds in France were authenticated. Unfortunately, the object in question was badly published, and disappeared from 1895 until its rediscovery in 1956! Consequently, very few works on Palaeolithic art mention it. This mineralised sacrum (base of spine) of an extinct fossil camelid was found at Tequixquiaq, in the northern part of the central basin of Mexico. The bone is carved and engraved – two nostrils have been cut into the end – so as to represent the head of a pig-like or dog-like animal; the circumstances of its discovery are unclear, but it is thought to be from a late Pleistocene bone bed, and to be at least 11,000 or 12,000 years old;[2] it is on exhibit in Mexico's National Museum of Anthropology.

Other examples of portable art in the New World are less well authenticated, although a bone with an engraving of a rhinoceros from Jacob's Cave, Missouri, is thought to be of Pleistocene date.[3] The mammoth engraved on a fossil shell pendant, found at Holly Oak, Delaware, in 1864 (Fig. 9) is still the subject of some doubt regarding both its Pleistocene date and its authenticity.[4]

As for parietal art, this is far more difficult to date. The New World has decorated caves and rock-shelters in many areas, such as southern Peru

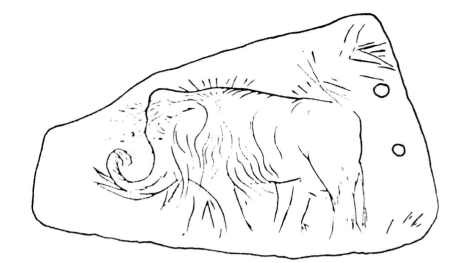

Fig.9   *Tracing of the mammoth engraved on a piece of a large whelk, from Holly Oak (Delaware). Authenticity uncertain. (After Kraft & Thomas)*

and Patagonia,[5] which some scholars believe to date back to the Pleistocene; they may be right, but proof is lacking for the present.

Recently, however, a layer in the decorated rock-shelter of Pedra Furada, Brazil, which is said to contain a painted fragment fallen from the wall, has been dated to 17,000 years ago;[6] new methods of dating layers of varnish formed on rocks are starting to suggest the possibility of very early origins for rock-carvings in the western part of North America – in eastern California, some layers formed on top of rock-engravings have been dated to the eleventh or twelfth millennium before the present.[7]

Africa

Once again, it is very difficult to date art in rock-shelters, but it is extremely likely that some examples in various parts of the continent are of late Pleistocene age – in Zimbabwe, fragments of painted stone are known from layers at least 13,000 back to more than 40,000 years old, and pigment has been in use for at least 125,000 years.[8] Portable Palaeolithic art has been well authenticated in South West Africa: seven fragments of stone found by Eric Wendt in the Apollo 11 cave, southern Namibia, have paint on them, including four or five recognisable animal figures such as a black rhino and two possible zebras; they display a use of two colours, and were associated with charcoal which has provided a radiocarbon date of at least 19,000 and perhaps even 26,000 years ago.[9] Engraved pieces of wood and bone (including a baboon fibula with 29 parallel, incised notches) have been recovered from Border Cave, Kwazulu, and dated to between 35,000 and 37,500 years ago, while a bored stone decorated with incisions has been found in a layer of Matupi Cave, Zaïre, dating to about 20,000 years ago.[10]

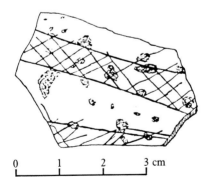

0   1   2   3 cm

Fig. 10   *Tracing of an engraved fragment of ostrich eggshell from Patne, India, dating to at least 25,000 years ago. (After Kumar et al)*

Arabia and India

It has been claimed that the oldest rock-art in central Arabia dates back some 14,000 years, though the only evidence for this is the apparently Pleistocene fauna depicted[11] – the same situation exists in India, where the hundreds of caves and rock-shelters around Bhimbetka, near Bhopal, contain parietal paintings spanning a long period: claims have been put forward that the earliest are Upper Palaeolithic in age, especially since engraved ostrich eggshells from excavated layers here are said to have been dated to between 25,000 and 40,000 years ago.[12]

The Far East

China's Palaeolithic 'art' is limited, for the moment, to beads and other decorative objects from the upper cave at Zhoukoudian. However, You Yuzhu is studying the animal bones from the important site of Shiyu, in Shanxi Province (NW China), which dates to 28,135 years ago and has yielded over 20,000 stone tools. He has found several hundred bones with cut-marks and 'engravings' on them, but so far only one has 'figures'. He believes that this specimen, a fragment of horse femur, shows an engraving of an animal approached by two hunters, and a large bird pursued by other hunters.[13] I have not seen the original bone, but no specialist to whom I have shown a photograph of the object has been able to see anything figurative in the marks, and it is not even clear that they were made by a tool. No doubt the future will bring better examples from China; for the moment, the problem remains open.

The same applies to Korea where a number of claims have been made for portable art in the Middle Palaeolithic: these take the form of bones supposedly modified to depict animals,[14] but serious doubts exist about all of them. Similarly, primitive pecked rock-carvings in a number of sites are claimed to be Palaeolithic, in part because they are thought to depict extinct reindeer and grey deer, and partly because they display 'Palaeolithic mentality',[15] but, once again, proof is lacking.

In Japan, on the other hand, there are some extremely interesting engraved pebbles from the cave of Kamikuroiwa. Layer IX (Initial Jomon) has been dated to 12,165 years ago, and contained several little pebbles with engravings on them, some of which seem to represent breasts and 'skirts' (Fig. 11).[16]

Australia

It is in Australia, with its incredible wealth of rock-art, that one finds most of the non-European examples of Pleistocene parietal art. The first site where its existence was authenticated was Koonalda Cave, in southern Australia, which was found to contain abundant 'digital flutings' (lines made with fingers) on the ceiling and walls, in total darkness, hundreds of metres inside; they seemed to be associated with the extraction of flint.[17] The site's archaeology showed that the mining activity took place at least 15,000 to 24,000 years ago, and so the marks are probably of similar age. These finger flutings are identical to those known in several of the European Palaeolithic caves.

Like those of Koonalda, finger flutings in the Snowy River Cave, Victoria, can be dated only by assuming that they are contemporaneous with archaeological deposits at the cave-entrance, which are about 20,000 years old.

The first direct proof of the antiquity of Australian art came from the Early Man shelter in Queensland, where very weathered and patinated engravings (circles, grids and intertwined lines) covered the back wall and disappeared into the archaeological layer (Fig 12). As the layer yielded a radiocarbon date of 13,000 years ago,[18] it is clear that the engravings must be at least this old: it will be recalled that it was precisely this sort of proof, at La Mouthe, which finally clinched the authenticity of parietal art in Europe.

Engravings resembling those of Early Man shelter exist in Tasmania; as that region became separated from the continent by a rise in sea-level

Fig.11 *Tracing of 'Venus' pebbles from Kamikuroiwa, Japan. Each about 5 cm long. (After Aikens & Higuchi)*

Fig.12 *Engravings in the Early Man shelter, Queensland. (Photo A. Rosenfeld)*

around 12,000 years ago, and since its art contains no engraved dingo prints (the dingo having arrived in Australia after that date), some scholars believe that the Tasmanian engravings are older than the separation; others think they are much younger. In 1986, 16 red hand stencils were found deep inside the Ballawine Cave in the Maxwell Valley, SW Tasmania; they are undatable, but, from a comparison of the cave's archaeological material with that of similar sites in the region, it has been estimated that they may be at least 14,000 years old.[19] In 1987, hand stencils and areas of wall smeared with red were found in another Tasmanian cave, Judd's Cavern.[20]

Arnhem Land, in northern Australia, is extremely rich in rock-art; ochre 'pencils' with traces of wear have been found there in layers dating to 18,000 and 19,000 years ago, and perhaps even 30,000. Some of the oldest paintings are covered by a thin siliceous film, which is deposited only in very arid conditions – the last such period in this region occurring 18,000 years ago. Moreover, among these apparently ancient figures are animals which have been interpreted as species extinct in Australia for at least that length of time (eg the marsupial tapir *Palorchestes*),[21] although some scholars disagree with these interpretations.

For the moment there is very little portable art of this period: in the cave of Devil's Lair, Western Australia, three perforated bone beads have been found in a layer that is 12,000–15,000 years old, together with a possible stone pendant, and some slabs of stone with engraved lines on them (though it is not certain that the lines are artificial).[22]

Figs. 13/14   *Large, deeply carved circle with internal lozenge lattice in entrance of Paroong Cave, S Australia, and deeply engraved circles (diameter: between 15 and 45 cm) on the opposite wall of the entrance.(PGB)*

In NW Australia, in the Pilbara, there are thousands of engravings on rocks in the open air, and many scholars are starting to think that the patina on some of them proves that they are extremely ancient, probably some 10,000–15,000 years old; and dating of 'desert varnish' (see New World, above) covering petroglyphs in the Olary region of South Australia recently produced a result of over 30,000 years![23]

There are not huge numbers of caves in Australia, and those which do exist are shunned by modern Aborigines, who are wary of the dark and of caverns. Consequently, any art found inside a cave is unlikely to be of recent origin. The discovery of digital flutings at Koonalda was followed by similar though lesser finds in the small Orchestra Shell Cave (Western Australia)[24] and in two other caves in the south of the country. Recently, however, a whole series of decorated caves has been discovered in South Australia, near Mount Gambier. This region contains hundreds of caves in its Tertiary limestone karst. Up to now, 140 have been examined; almost all contain traces of animals (clawmarks on the walls, and so forth), and 22

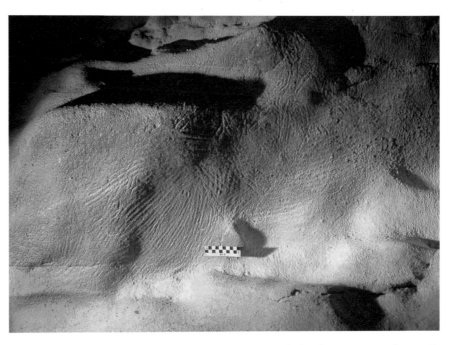

Fig. 15 *Digital tracings in Karlie-ngoinpool Cave, S Australia. (PGB)*

of them also contain non-figurative marks made by humans on the walls and ceilings. The discoverer, Geoff Aslin, and his collaborators, Robert and Elfriede Bednarik, have undertaken an in-depth investigation of what seems to be one of the most remarkable concentrations of non-figurative parietal art in the world.

These caves are not like the well-known examples in France and Spain; many of them have vertical entrances, at ground level in fields. Farmers have always used them as garbage dumps, throwing in scrap metal, machines and all kinds of refuse, to prevent livestock falling in; consequently the descent into the depths can be dangerous, as one negotiates a tottering mountain of rusting barbed wire and old refrigerators. This state of affairs, however, has helped to protect the caves and their art: very few potholers or explorers have entered them, especially

Fig. 16 *Deeply engraved circles associated with extraction of flint nodules. Karlie-ngoinpool Cave, S Australia. (PGB)*

as they are on private land and often near the farmhouses. These caves are to be cleared of rubbish and locked, one by one: until then, their exact location must remain a secret.

It has been established[25] that they contain three traditions of petroglyphs, stylistically distinct, which are sometimes superimposed on the same wall, separated by thin layers of carbonate. It is hoped that radiometric dating (the Uranium/Thorium method) of these deposits will one day provide an approximate chronology for the traditions.

The earliest phase comprises the digital flutings; all 22 caves have them, and they always lie beneath the engraved lines and seem to follow or emphasise the topography of the walls. Like those of Koonalda, Orchestra Shell Cave and others, they are probably at least 20,000 years old. At that time, therefore, this tradition of finger-marking extended along the entire southern part of the continent, a distance of 3,000 km!

The second phase, also very ancient, is called Karake (after the cave where such traces were first seen); it is characterised by deeply engraved and weathered circles – simple, concentric, or divided up by gouged lines – which closely resemble the archaic petroglyphs of Early Man shelter (13,000 years old) and the Tasmanian engravings. In other words, this tradition seems to have extended down the eastern side of the continent.

Finally, the terminal phase, comprising shallow engravings, is probably much more recent, dating to the Holocene period.

It is noteworthy that several of these caves, like Koonalda, served as chert mines at the time of their decoration; the region around Mount Gambier was already known to archaeologists for the great abundance of Pleistocene stone tools and knapping-waste found there, around the good-quality surface sources of stone. The decorated cave-mines are thus only the nucleus of a considerable activity area.

Two of the caves in particular are outstanding. Paroong has an important collection of Karake-type petroglyphs, especially at its entrance: the vertical walls here are literally covered with a profusion of deeply carved circles, one next to another, up to a height of 4 m above the present floor (Figs. 13,14).

The cave of Karlie-ngoinpool (which means 'many' in the old local indigenous language) contains one of the greatest known concentrations of non-figurative parietal art. Its well-preserved finger flutings cover at least 75 square metres (Fig. 15). There are also quantities of deeply engraved circles and other Karake motifs, and some of them (covering 10 m width of wall) remain hidden behind the mountain of rubbish in the entrance. Many of these circles are closely associated with the extraction of nodules of chert of very poor quality (Fig. 16). The cave ends in a little 'sanctuary', a tiny chamber in total darkness and covered in engravings, especially circles, which appears to have been of particular significance for the people of this remote period.

Clearly, therefore, the map of Pleistocene artistic activity is rapidly filling up, with Australia providing a number of well-dated examples. It is probable that both that country and many others will have innumerable further surprises of this type in store for us. Nevertheless, Europe remains supreme, for the present, in the quantity and quality of its surviving Palaeolithic art.

Europe

The Palaeolithic pictures of Europe are generally treated as two distinct entities, the portable and the parietal – whereas in reality these are merely the two ends of a continuous range; in other words, there is an overlap between the two categories, comprising cases in which it is impossible to decide whether detached fragments of wall were decorated before or after falling, and blocks which could be moved but were too large to be carried around (Breuil called them *parois mobiles*, or movable walls): at the French rock-shelter of La Marche, for example, the size of the engraved slabs varies from specimens a few centimetres long to some over a square metre in area and weighing tens of kilograms. However, for the sake of convenience, the usual artificial division will be retained here.

Portable

The two categories have somewhat different distributions within Europe. Portable art is found from the Iberian Peninsula and North Africa to Siberia,[26] and has notable concentrations in Central and Eastern Europe; occasional specimens also turn up in countries around the fringes of Europe, such as the crude engraving of a possible animal on a limestone slab from the Aurignacian site of Hayonim Cave, Israel,[27] or the handful of Upper Palaeolithic engraved bones from England.[28]

It is very hard to quantify portable art objects, since many are broken, and unknown numbers of them remain unpublished, whether in private or clandestine collections, or lying forgotten and unstudied in museums around the world.[29] One estimate in 1980[30] that there are well over 10,000 pieces of Palaeolithic portable art in Western Europe alone is certainly a minimal figure. A 1965 study[31] of the portable art from sites in the Périgord region of France produced a total of 2,329 objects – once again a minimal figure, since there are many more in foreign or private museums and collections.

The scattering of these objects around the museums of the world poses great problems for their study, which will be alleviated only when good casts of the 'absentees' can be housed in some European research centres, and when full details of every specimen are available from the computerised data banks which are now being compiled.[32]

On the other hand, the uneven distribution of the objects among archaeological sites is a stimulating puzzle for the archaeologist. Many Palaeolithic sites in Europe have no decorated objects at all, while others have only one or a few. Some Magdalenian 'supersites', however, have hundreds: the cave of Parpalló, in eastern Spain, produced over 5,000 plaques of engraved and painted stone, together with scores of decorated bones, from its long Upper Palaeolithic occupation, and especially the Magdalenian.[33] The site of La Marche, already mentioned above, has at least 1,512 engraved slabs of stone (1,268 after some fragments were fitted together), weighing a total of 4 tons![34] The open-air site of Gönnersdorf in West Germany has about 500 engraved slates.[35] The cave of Enlène, in the French Pyrenees, has recently yielded over 1,000 engraved stones, with more being found every year, in addition to its wealth of decorated bone and antler[36] – in some areas of the cave floor, over 100 engraved slabs have been found in less than a square metre!

Clearly, therefore, some sites are infinitely more important than others in terms of decorated objects. This becomes particularly clear from the

above-mentioned survey of Périgord specimens: the 2,329 items were spread over 72 sites; but two rock-shelters alone, La Madeleine (582 objects) and Laugerie-Basse (560), account for 1,142, or 48% of the total; if one adds Limeuil (176) and Rochereil (132), then only four sites account for 1,450 objects, or 66% of the total. Similar figures would emerge from studies of this kind in the Pyrenees or Spain and, as we shall now see, parietal art's distribution is equally patchy.

Parietal

Cave-art is difficult to quantify, both in numbers of caves and in numbers of pictures. The most striking fact, however, is that its geographical

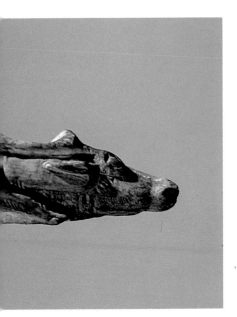

Fig. 17 *Ivory carving depicting two reindeer, perhaps a male following a female, from Bruniquel (Tarn et Garonne). Probably Magdalenian. Length: 20.7 cm. (JV)*

Fig. 18 *Antler spear-thrower carved in the form of a mammoth, from Bruniquel (Tarn et Garonne). The trunk joins the legs while the tusks jut forwards. The ears are depicted near the base of the tusks, but the eyes are absent (the hole is unlikely to represent the eye since it is not in the correct position). Probably Magdalenian. Length: 12.4 cm. (JV)*

distribution in Europe is very different from that of portable art, although it is most abundant in areas which are also rich in decorated objects: the Périgord, the French Pyrenees and Cantabrian Spain.

Palaeolithic decorated caves are found from Portugal and the very south of Spain up to the north of France: the most northerly known at present is that of Gouy, just south-east of Rouen.[37] None has yet been found in Belgium or Britain,[38] though that may change in the future. They do not extend to the east of France, except along the Mediterranean through Sicily and Italy to western Yugoslavia and Romania.

The first problem in quantifying decorated caves is that they cannot usually be dated directly (see below, p.54) and therefore many examples remain doubtful, especially where they contain only non-figurative designs. For example, claims appear sporadically for parietal art in countries such as Hungary and Anatolia, but all are lacking in proof at present. For a while, an engraving of a stag in the Bavarian cave of Schulerloch was thought to be Palaeolithic,[39] but, together with another German candidate, it is now thought to be of more recent date.[40]

The Soviet Union has a number of these doubtful cases from Georgia to Siberia, in caves, rock-shelters and on open-air cliff-faces; they comprise both engravings and paintings, and both realistic and stylised animals and geometric motifs.[41] Their possible antiquity received a boost in 1959 when what seem to be genuinely Palaeolithic paintings were discovered in the great cave of Kapova (or Kapovaya, or Shulgan-Tash) in the southern Ural Mountains; the red figures include eight mammoths, two rhinos and two horses.[42] The cave of Ignatiev, 100 km distant in the southern Urals, has also been accepted as having Palaeolithic parietal art, including an outline drawing of a horse.[43]

The fact that these caves are 4,000 km away from the main clusters of decorated caves in western Europe serves to highlight the difference in distribution between portable and parietal art. No decorated caves have yet been found through all of Central Europe, between north-east France and the Urals, despite an abundance of portable art across this void. Of course, cave-art can occur only in areas with caves (though, as we shall see below, p.110, examples of open-air rock-art are now being found); but Central Europe has plenty of suitable sites. Some theories about the discrepancy will be examined later.

It should not be thought, however, that all decorated caves in western Europe are well authenticated; apart from occasional suspected fakes (also a problem with portable art), there are many cases in which reported examples of cave art have never been rediscovered; have been proved to be natural shapes, cracks or mineral colours in the rock (*lusus naturae*); have been deemed post-Palaeolithic; or simply remain doubtful.[44] This means that no two tallies of decorated caves are likely to be identical: in the most recent authoritative survey of French cave-art, no fewer than 42 examples were considered erroneous or very doubtful in the Aquitaine region alone.[45] Another problem with quantification is that some caves which are now separate may have formed a single system in the Upper Palaeolithic; while other caves have been given several names which become confused in the literature, so that the same site acquires two or three different entries in the list!

According to the most recent survey (1984), there are 130 accepted and authenticated examples of Palaeolithic parietal art in France,[46] although at least seven more have been discovered since then; there are 106 examples

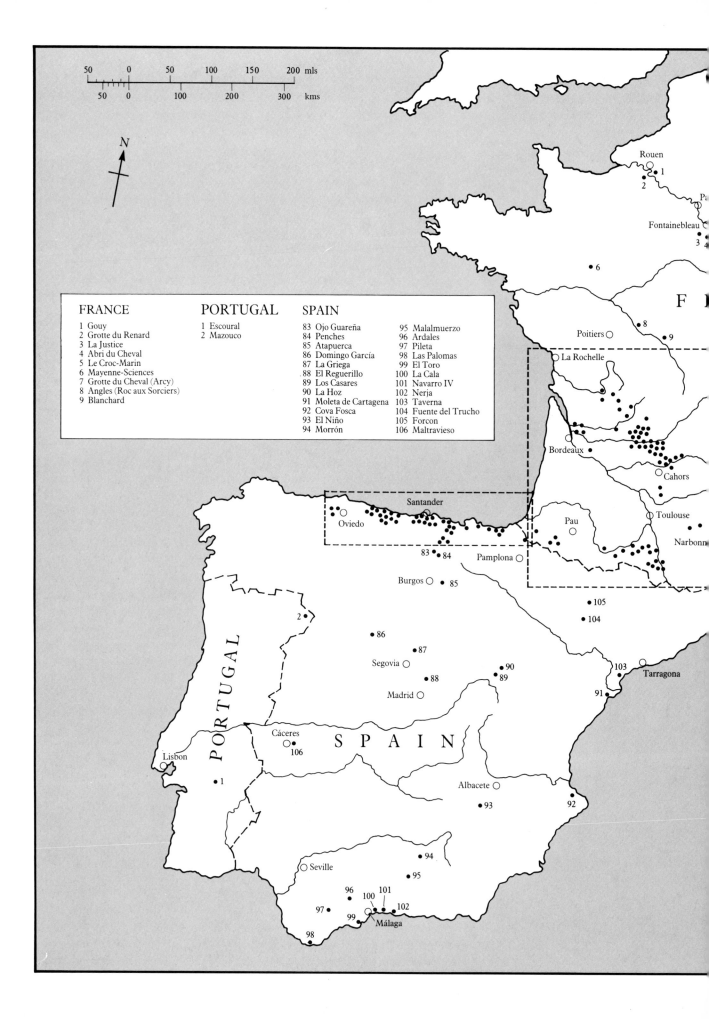

FRANCE

1 Gouy
2 Grotte du Renard
3 La Justice
4 Abri du Cheval
5 Le Croc-Marin
6 Mayenne-Sciences
7 Grotte du Cheval (Arcy)
8 Angles (Roc aux Sorciers)
9 Blanchard

PORTUGAL

1 Escoural
2 Mazouco

SPAIN

83 Ojo Guareña
84 Penches
85 Atapuerca
86 Domingo García
87 La Griega
88 El Reguerillo
89 Los Casares
90 La Hoz
91 Moleta de Cartagena
92 Cova Fosca
93 El Niño
94 Morrón

95 Malalmuerzo
96 Ardales
97 Pileta
98 Las Palomas
99 El Toro
100 La Cala
101 Navarro IV
102 Nerja
103 Taverna
104 Fuente del Trucho
105 Forcon
106 Maltravieso

Map of Europe showing approximate locations of parietal art sites

Kapova & Ignatiev

Cuciulat

YUGOSLAVIA

N C E

Lyon

Como 4

Torino

Genova

Nice 3

Marseille

ITALY

22

Rome

12

Lecce
16

17

Cosenza

ITALY · SICILY
YUGOSLAVIA

1 Addaura
2 Armetta
3 Balzi Rossi Group
4 Buco della Sabia di Civate
5 Giglio
6 Giumente
7 Isolidda group
8 Levanzo
9 Miceli
10 Santa Rosalia group
11 Niscemi
12 Paglicci
13 Pizzo Muletta
14 Puntali
15 Roca Rumeni
16 Romanelli
17 Romito
18 Sallinella
19 San Teodoro
20 Vaccari
21 Za Minica I/II
22 Badanj

14
15 7 10
8   13 1 11
9   21 2   6
5   20
18   Palermo

19

SICILY

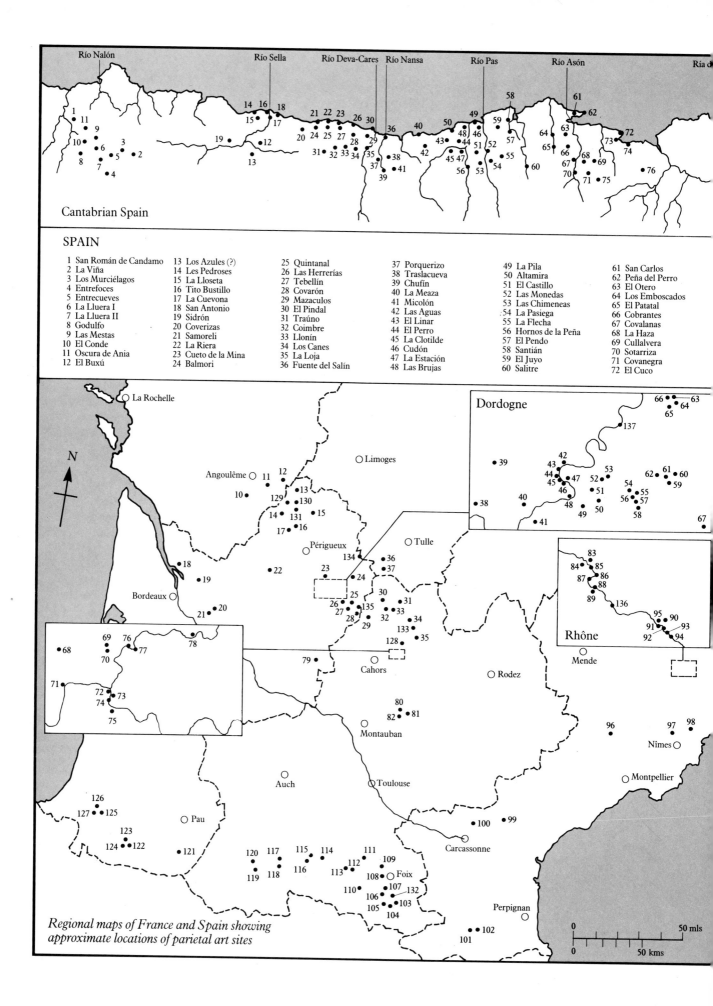

Río Nalón Río Sella Río Deva-Cares Río Nansa Río Pas Río Asón Ría d

Cantabrian Spain

SPAIN

1 San Román de Candamo
2 La Viña
3 Los Murciélagos
4 Entrefoces
5 Entrecueves
6 La Lluera I
7 La Lluera II
8 Godulfo
9 Las Mestas
10 El Conde
11 Oscura de Ania
12 El Buxú

13 Los Azules (?)
14 Les Pedroses
15 La Lloseta
16 Tito Bustillo
17 La Cuevona
18 San Antonio
19 Sidrón
20 Coverizas
21 Samoreli
22 La Riera
23 Cueto de la Mina
24 Balmori

25 Quintanal
26 Las Herrerías
27 Tebellín
28 Covarón
29 Mazaculos
30 El Pindal
31 Traúno
32 Coimbre
33 Llonín
34 Los Canes
35 La Loja
36 Fuente del Salín

37 Porquerizo
38 Traslacueva
39 Chufín
40 La Meaza
41 Micolón
42 Las Aguas
43 El Linar
44 El Perro
45 La Clotilde
46 Cudón
47 La Estación
48 Las Brujas

49 La Pila
50 Altamira
51 El Castillo
52 Las Monedas
53 Las Chimeneas
54 La Pasiega
55 La Flecha
56 Hornos de la Peña
57 El Pendo
58 Santián
59 El Juyo
60 Salitre

61 San Carlos
62 Peña del Perro
63 El Otero
64 Los Emboscados
65 El Patatal
66 Cobrantes
67 Covalanas
68 La Haza
69 Cullalvera
70 Sotarriza
71 Covanegra
72 El Cuco

Dordogne

Rhône

Regional maps of France and Spain showing approximate locations of parietal art sites

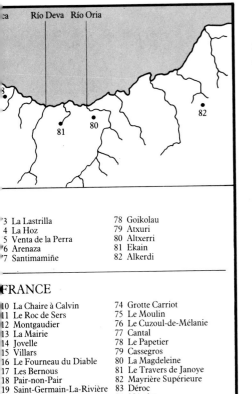

in Spain,[47] up to 21 caves/groups in Sicily/Italy,[48] 2 sites in Portugal (Escoural and Mazouco), 1 in Yugoslavia (Badanj),[49] and 1 in Romania (Cuciulat).[50] As has already been mentioned, their distribution within countries is very uneven, with important clusters occurring both in the Périgord (48 sites), in the French Pyrenees (29 sites) and in Cantabrian Spain (82 sites from Asturias to Navarre).[51]

Moreover, just as portable art varies markedly in quantity from site to site, the same is true of parietal images. As mentioned above, it is difficult to quantify the drawings: this is because of superimpositions, deterioration and, above all, a lack of consensus about how to count 'signs' – should a series of dots be considered separately or as a unit? And how can one quantify a mass of meandering fingermarks? Consequently, different authors often have widely divergent totals of figures for the same cave.

Nevertheless, it is clear that there are some sites with few figures and others with hundreds. For example, the Grotte du Roc at Saint-André-d'Allas (Dordogne), or the Cueva de los Murciélagos at Portazgo and the Abrigo de Godulfo at Bercio (both in Asturias) have one figure apiece;[52] others, like the Grotte de Pradières (Ariège) or the Cova Bastera (Pyrénées Orientales) have only a few red marks on their walls;[53] while the Grotte du Cheval at Foix (Ariège) or the Abri du Poisson (Dordogne) have a mere handful of figures.[54]

On the other hand, Lascaux (Dordogne) has about 600 paintings and nearly 1,500 engravings;[55] the Grotte des Trois Frères (Ariège) has over 1,100 parietal figures,[56] more than any other Pyrenean site – indeed, since that figure was calculated, whole new areas of engraving have been found by careful removal of clay from the walls, and more remain covered for the moment;[57] some Spanish caves, such as Castillo and Altamira, are equally dominant in their region. It is very hard to see how caves with one figure and supersites like these could be equivalent in any way; similarly, some caves are huge while others are tiny, though size does not always equate with numbers of figures – there are large caves with few, and small caves with many.

In an exercise similar to that on the Périgord's portable art, the 'graphic units' of the parietal sites in Ariège (and part of Haute Garonne) have been quantified: the total of over 2,600 'graphic units' is spread over 11 caves, but four of them (Trois Frères, Niaux, Fontanet, Marsoulas) account for 2,216 units, or 84.9%,[58] with Trois Frères predominant of course (over 50%).

Both types of art

Thus we have some sites with huge quantities of portable art, and some with similar concentrations of parietal images. Do the two phenomena ever coincide? Surprisingly, in those regions of Europe where the two types of art are found, there are relatively few sites which have both (and many which have neither form!): a recent study of this topic, focusing on figurative art, found only 27 sites in France, 9 in Spain and 3 in Italy[59] – and some of these had only one portable item, while others had over 100. Some local areas rich in one form are noticeably poor in the other.

It seems reasonable to suppose that caves which have both art-types, with one or both in large quantities, were very special places in the last Ice Age: one can certainly include the great river tunnel of Le Mas d'Azil (Ariège) and Isturitz with its three decorated tunnel-caves – these are the true 'supersites' of the Pyrenees, with their huge concentrations of tool-production and artistic activity; they must have served as storehouses, meeting-places, ritual foci and socio-economic centres not only for local groups but also for a far wider area, as is confirmed by their artefactual links with far-flung sites and with each other.[60] The same is true, though on a lesser scale, of Altamira and Castillo.[61]

However, perhaps the most dramatic example can be seen in the Volp caves: Enlène (which, as we have seen, has huge amounts of portable art) is connected by a passage to Trois Frères, the richest parietal site in the whole region. This cave-system, therefore, was clearly of the greatest importance in the Magdalenian period, an importance further underlined by the immediate proximity of the remarkable Tuc d'Audoubert (see p.94).

Enlène was inhabited; it has hearths, but no wall art. Trois Frères is profusely decorated, but has almost no trace of habitation. These caves thus suggested to early scholars of Palaeolithic art that living sites were never decorated, and 'sanctuaries' were never inhabited. Their view seemed to be further strengthened by the case of Niaux and La Vache, two caves facing each other across a narrow valley in Ariège. Niaux is rich in wall-art but, as far as we know, was not inhabited; La Vache is a living site, rich in portable art but with no Palaeolithic wall decoration. However, it is now known that some decorated sites *were* inhabited (eg Lascaux, Fontanet, Marsoulas), though in some cases it may still be true that a decorated site was visited rarely, or only once, for whatever rites were carried out there.

As will be discussed later (p.113), there are clear regional and chronological differences in the techniques and content of both portable and parietal art in Europe; taken as a whole, however, this phenomenon now comprises about 275 decorated sites and many thousands of objects. More are being found every year – the objects mostly by archaeologists, and the caves (an average of about one per year) primarily by spelaeologists and quarrymen. It may seem a lot but, seen against the 25,000 years of the Upper Palaeolithic, or even just the few millennia of the Magdalenian, the known quantity of Palaeolithic art actually remains quite small. One might say that as a medium it is rare, but often well done...

3
Making a Record

Tracing and copying

Since the first discoveries and the acceptance of Palaeolithic art, a great deal of effort has gone into copying and recording it in one way or another. This is primarily to make reproductions available for scholars to work with (many examples need to be studied for research or synthesis), and to present the material to the public at large. Copying has the additional advantage of reducing the number of occasions on which an original object might need to be handled or a cave visited – and in a few cases even specialists cannot visit the caves at present: the Réseau Clastres at Niaux is closed off by an underground lake and kept sealed in this way so that instruments inside it can send out data on the cave's undisturbed 'climate'; while Erberua (Pyrénées Atlantiques) lies at the end of a long 'siphon', and can thus be visited only by experienced divers with modern equipment.

Finally, of course, the making of copies ensures that any example of Palaeolithic art which might deteriorate or be lost, stolen or damaged will at least survive in reproduction: for example, one of the Laussel 'Venus' bas-reliefs is thought to have been destroyed in Berlin during the last war, and survives only in the form of casts and pictures.[1] Photographs and tracings made over the last 100 years also enable us to monitor changes to the original images through time.[2]

Fig. 19   *The young Henri Breuil (1877-1961). (Photo PGB collection)*

Breuil

The doyen of recorders of the art was the abbé Henri Breuil, who was a young man at the turn of the century when the first decorated caves were authenticated (Fig. 19): he was a talented artist, excelling in animal figures, and had been employed by Edouard Piette to draw his remarkable collection of Palaeolithic portable art.[3] He therefore had the great good fortune to be present – young and vigorous, with time on his hands and possessing the necessary skills – at the very time when the copying of cave-art needed to begin. His first attempt entailed tracing a few figures in La Mouthe in 1900;[4] subsequently, he did Altamira in 1902, Marsoulas, Font de Gaume, Combarelles, and all the other early finds – indeed he himself found many caves or figures.

His working methods were very crude by modern standards, though he was hampered by the limited materials then in existence. Sheets of florist's or rice paper (the most transparent available, though merely translucent at best) were held on to the cave-wall, often by assistants, and the Palaeolithic lines beneath were traced with pencil or crayon. Carbide lamps had to be held, often at arm's length for long periods, by helpers; Breuil was quick-tempered, and might administer a sharp slap if a boy's fatigue or cramp caused the light or the paper to move;[5] he was utterly absorbed in the work, hardly speaking for hours on end, simply moving the arms of his 'human candelabras' when necessary.[6]

Later the tracings would be redrawn for publication, though not always immediately – Breuil did so much copying, and had so many other commitments, that publication sometimes occurred many years later (20 years in the case of Trois Frères, 24 for the Galerie Vidal at Bédeilhac, 26 for Pair-non-Pair), and the redrawing may have been done by a collaborator (eg by Michaut in the case of the Galerie Vidal), and then published unchecked. This inevitably led to mistakes: for example, his published tracings of Pair-non-Pair contain numerous meanders and circular images, but recent investigation has shown that these are all natural accidents in the rock-face: fissures, bumps and hollows; he undoubtedly noted them as such on his original tracings, but when the time came for publication the meaning of his graphic conventions had been forgotten.[7]

The 'direct tracing' method had its disadvantages, since contact with the wall inevitably damaged the art very slightly in some places (as can be seen in some modern macrophotographs), and Breuil was even known on rare occasions to dust surfaces with a few flicks of his handkerchief to make some lines clearer![8] Even fragile engravings in the clay of cave-floors, such as those at Niaux, were traced by the direct method, running a pencil without pressure on to soft paper in contact with the clay.[9]

At Altamira, his working conditions in 1902 were appalling: since the ceiling figures were done in a very pasty paint, his paper could not be placed on them, as this would detach pigment. Consequently, he had to copy the animals while lying supine on sacks filled with ferns, using the imperfect yellow light of candles which spattered his clothing with wax, and undergoing constant fatigue and strain. His method in this cave was to make rough sketches, then measure the originals, and finally make his copies. He had brought a box of water-colour paints with him, but the cave's atmosphere was too humid for the paper to dry, so he used pastels instead; having no black (a colour of great importance in the cave), he had to improvise with burnt wood and crushed charcoal mixed with water, but the results were unsatisfactory; he returned to Altamira in 1932 to do the job properly, redrawing a few figures and retouching the old copies of the rest.[10] Not only was electric light now available; the floor had been lowered, which meant that he could be comfortably seated and could see the figures with less distortion and foreshortening than had been the case in 1902. It is worth noting that a few figures had already shown marked deterioration in the 30 intervening years.

Students of Palaeolithic art owe a huge debt of gratitude to Breuil for his patient toil in the caves – by his own reckoning he spent over 700 days deciphering and copying their art during the course of his long life. There

Fig. 20 *'Spot the Breuil': his version and a more recent tracing of an ibex engraving from Castillo (Santander): about 40 cm long. (After Almagro)*

Fig. 21 *'Spot the Breuil': his version and a more recent tracing of a horse engraving from Trois Frères (Ariège): 27.5 cm long. (After Bégouën & Clottes)*

are a number of cases where paintings have faded since he copied them (eg the Galerie Vidal in the cave of Bédeilhac), so that only the Breuil versions give one any idea as to how the originals may have looked.[11] But a major problem – which has come to the fore only in the last few decades – is that all his copies are indeed a 'Breuil version'. His colossal output (including his drawings of hundreds of portable items), his influence,[12] and his sheer dominance (not only in Palaeolithic art but in all prehistory) until his death in 1961, have ensured that Breuil copies are to be found in most textbooks and in most works on cave-art; in many cases they are far more familiar to us than the originals, and this is a very dangerous state of affairs because, like every artist, Breuil had a style of his own. We are therefore seeing Palaeolithic figures that have passed through a standard 'Breuil process': they are subjective copies, not faithful facsimiles. It is probable that a scholar of Ice Age art, shown six versions of the same figure by different prehistorians, could 'spot the Breuil' at a glance (Figs. 20,21).

There are three important consequences of this state of affairs. First, the Breuil style has inevitably made a wide range of Palaeolithic figures seem somewhat similar to our eyes; it may therefore exaggerate our view of the

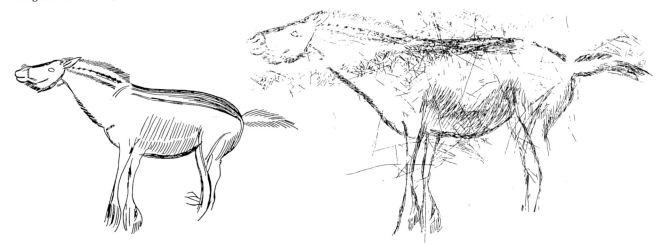

art's unity through space and time (and, conversely, comparison of figures produced by two different copiers may exaggerate the differences, since their own styles enhance the contrast).[13]

Second, Breuil's tracings are so well known and so firmly established that they can hamper new readings of a figure, since they influence what the researcher sees.[14]

Third, Breuil's copies reflect his own preoccupations, his own notions about the art. As will be seen (p.151), he believed in ideas of hunting magic: this was a hunting art, and he therefore went all out to decipher animal outlines, sometimes filling in missing elements. According to some accounts, he was fairly objective in his copying: while working in Trois Frères one day, he was asked the meaning of the particular line he was tracing; 'I don't know,' he replied, 'I copy what there is, we'll explain it later ... if it's possible'.[15]

Nevertheless, it is clear from his books and articles that, despite this attitude to his task, he was in fact extremely selective in the lines which he chose to copy or at least 'extract' for publication. Where lines seemed to have no relevance to the animal figures, he often ignored them, dismissing them as *traits parasites*'. As a result, we have lots of nice clean tracings of animals, copied from book to book and giving the impression that Palaeolithic art contains nothing else; whereas in fact there is a vast amount of non-figurative, abstract or geometric marking in the caves which may have been just as important, if not more important, to prehistoric people.

Modern methods of copying

Since Breuil was the first to do this kind of work, he had to learn as he went along; and in view of this lack of precedent, and the materials available to him, it is scarcely surprising that he made mistakes in his methods and his results; indeed the wonder is that he did not make far more, and modern scholars are unanimous in their admiration for the tremendous quality and perception of his work.

Today, conditions have greatly improved, not only in terms of light-sources, but also of the truly transparent, supple plastics and acetates available to draw on, and new types of pens and markers. Nevertheless, direct tracing is now (theoretically) taboo, since we are more aware of the damage it can cause. Instead, 'tracing at a distance' is done, with the sheet set up in front of the wall; and some new techniques are also available: in the past, moulds made of parietal engravings (as at Pair-non-Pair) had to be of plaster or clay, and the risk to the originals was enormous. Rubber latex was a significant improvement, but today casts can safely be made of engravings on hard surfaces using elastomer silicones and polyesters which not only cause no damage but also are quick and easy to apply, and produce far more precise and resistant results.[16]

Indeed, the results are entirely faithful, exact replicas of the original in size and volume. Moulds have the added advantage that they turn parietal art into portable; they can be removed for study elsewhere under different lighting systems, and can eventually be displayed to the public. Casts can even be treated so that the engraved lines stand out far more clearly than on the wall: water mixed with ink is spread over the cast; when it is wiped away, some remains inside the engraved lines. This extra clarity has enabled researchers to find no fewer than 25 new engraved figures on a cast from Trois Frères cave.[17]

Fig. 22   *Tracing of the plasticine
imprint of an engraved horsehead
from La Marche (Vienne).
Magdalenian. Length from muzzle to
ear: 6.3 cm. Note the probable
halter. (After Pales & de St Péreuse)*

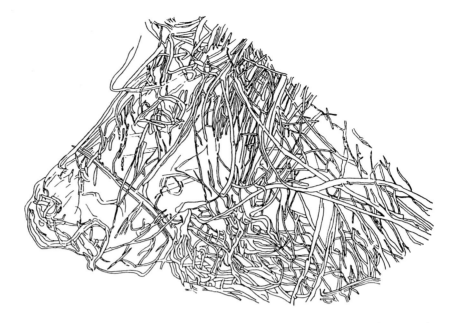

A similar technique has been applied to portable engravings such as those of La Marche and La Colombière. Plasticine or silicone imprints of the engraved surface turn the incisions into raised lines, and in some cases one can see in which order they were made. The imprints are also easier on the eye, since they do not have the distracting nuances of colour and texture of the originals.[18]

The study of portable art has also made great strides through use of the microscope, particularly in the pioneering work of Alexander Marshack;[19] this new methodology has revealed hitherto unnoticed details of content and composition and, as we shall see (p.75), can even suggest exactly how the different marks were made. A recent approach, combining casts and microscopes, involves making varnish replicas of the engraved surfaces of pebbles; these replicas, unlike the stones themselves, can then be studied under the Scanning Electron Microscope, at enormous magnifications; this method not only provides data on the engraving technique (see p.76), but can also show in what order superimposed lines have been engraved.[20]

Photography

All these new techniques are used in close association with photography. Thanks to the dominance of Breuil's tracings, little attention was paid to photographs in the field of Palaeolithic art until comparatively recently, despite some excellent early efforts: pictures were taken of the Altamira figures in 1880 using electric light; from October 1912 onwards, Max Bégouën was achieving extraordinary results in the Volp caves. He took the voluminous family camera to the farthest depths of the caverns; its glass negatives required poses of an hour, or quantities of magnesium, and were developed on the spot with water from the cave! He worked on them to produce maximum contrast, and the results thus give the false impression that the engravings have lost visibility since that time.[21]

Nowadays, thanks to the major advances in cameras, films and lighting, a wide range of techniques has been applied in the caves (see Appendix). For example, photographs of panels or figures are enlarged to full size; tracings are then made from these, though it is always necessary to check

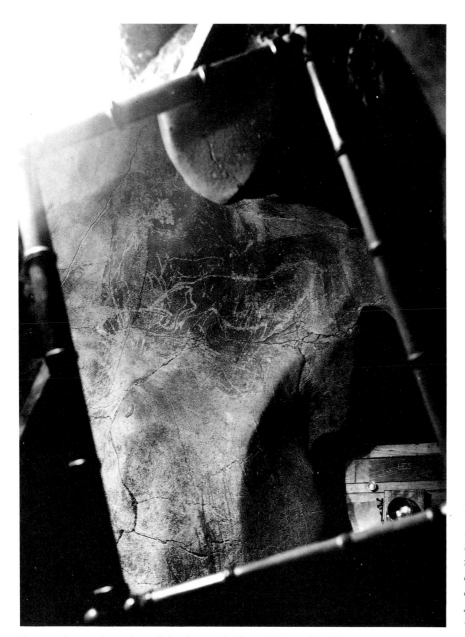

Fig. 23   *Pioneering photography in Trois Frères (Ariège) in the late 1920s. The cramped conditions in this gallery led to the use of the mirror. Note the camera's reflection and the horse engravings. (Photo courtesy of the Field Museum of Natural History, Chicago, negative #62007)*

the results against the original too – in fact, it is preferable to do the work in the cave, in front of the original. This technique is superior to other tracing methods, since the wall need never be touched, and one avoids the difficulties of installing the paper parallel to the panel, and of reflections and shadows caused by irregularities in the wall's surface.

Enlargements to greater than actual size, or macrophotographs of small areas, allow one to see tiny details or superimpositions more clearly. Photographs of paintings, of course, may be published as they are, but engravings often require tracings to help the reader decipher what is there: thus photography and drawing are complementary, not substitutes for one another.

By and large, photographs are taken perpendicular to the wall; for large figures or panels, or for those in narrow places where one cannot stand far enough back, a number of overlapping pictures are taken and then amalgamated into a photo-montage (Figs.30-2). The same technique is also invaluable for parietal figures which cannot be seen all at once (such as

Fig. 24   *Ultraviolet photo of red*
'claviform' sign in Trois Frères
(Ariège) painted on a surface
prepared by scraping. Probably
Magdalenian. The claviform is 76 cm
in length. (JV, collection Bégouën)

the horse 'falling' round a rock at Lascaux) or for portable objects such as those where the composition is engraved around a cylindrical baton.

A single picture of a parietal figure is far from sufficient today: a whole series is now taken, using different films (both monochrome and colour), light-sources (lamps and electronic flash) from different angles, lenses (normal, wide-angle, macro, etc), filters and degrees of contrast. Multiple flashes can bring out the relief of a sculpture, and even the traces of modelling clay by hand, as on the Tuc d'Audoubert bison.[22] The next few years will see the introduction of computers for the enhancement of photos and the storage of copies.

Another new technique, pioneered in the caves by Alexander Marshack and adopted by Jean Vertut and others, involves the use of infra-red and ultraviolet lamps. Ultraviolet radiation makes certain materials fluoresce: any calcite and living organisms on the cave-walls do so, but the ochres and manganese used as pigments do not; consequently one can assess any damage to the figures caused by growths or calcite-flows, while U/V can also show up any paint beneath the calcite and thus 'restore' fading detail, as shown by Marshack on some of the figures at Niaux.[23] Consequently, U/V lamps were used for both tracing and photography during the recent systematic recording of this cave's art.[24]

On the other hand, infra-red light or film makes red ochres appear transparent, so that one can see other pigments beneath them; moreover, any impurities in the ochres remain visible, and so different mixes of paint, with different impurities, can be detected. By this method, Marshack claims to be able to assess in what order the famous 'spotted horses' panel at Pech Merle (Lot) was built up.[25]

Infra-red may make the original composition clearer by 'removing' the thin trickles where pigment ran: Jean Vertut's picture of a 'tectiform' sign at Bernifal reveals that it was not painted in continuous lines, as some had thought, but as a series of dots from which the colour had run (Fig.26).[26] The use of infra-red also counteracts the effects of changes in humidity and wall-conditions, which can make certain painted lines visible on some days and not on others – for parietal paintings 'live' in accordance with atmospheric conditions.[27]

In addition to the traditional tracings, therefore, it has become essential for archives to include complete coverage in slide form (Jean Vertut took about 1,200 slides in the Volp caves), and in some cases also on video and film – as, for example, in the 'Corpus Lascaux', a cinematic archive of that cave's decoration.[28]

However, for art of this kind, the ideal is to have a record in three dimensions; so the very expensive technique of photogrammetry (the method by which contour maps are made from aerial photographs) is increasingly being used; in Palaeolithic art, by means of stereophotographs and some key measurements, it is possible to produce a detailed 'contour map' of an object or panel, which can then be used to make an accurate 3-D copy – in other words, making a cast without even touching the original! This has been done, for example, for the clay bison of the Tuc d'Audoubert, which could not be copied by any casting technique owing to their fragility[29] – Jean Vertut's expertise was invaluable in making this project a success.

The best-known 3-D copy at present is the excellent Lascaux II, completed in 1983, which reproduces part of that cave and its paintings;[30] but the technology which produced it is already almost obsolete, and even more accurate copies of decorated caves can and will be made in the future.

The quest for objectivity

Every tracing or copy of a Palaeolithic figure is inevitably subjective to some degree; it is also a distortion of reality since – except in the case of accurate casts – three dimensions are being reduced to two. Every tracing is a personal piece of work, and it is impossible to remove subjectivity completely. A copy is only as good as the copier, and all copiers make mistakes, the number and extent of which depend on the method they use and on their experience and personality: good examples include a series of 18 published versions of an engraved human head from the Grotte du Placard (Charente), none of which proved to be completely reliable;[31] five different versions of an engraved reindeer from Les Combarelles;[32] nine of a reindeer from Saut-du-Perron and seven of a rhino from La Colombière;[33] eight of the chamois on a disc from Laugerie-Basse;[34] and about 50 of the famous mammoth of La Madeleine, found in 1864.[35] Some of them were clearly done from the original, others were simply copied from book to book, with the distortions increasing each time. If the original were destroyed or lost, which of the versions should one 'believe'?

There are even a few cases where a truly appalling copy has been published then republished by others, who neglected to compare it with the original object (or even with a cast or photograph), although a glance would have sufficed to reveal the numerous errors.[36]

A great deal of emphasis is now placed on the 'depersonalisation' of copying; the pendulum has swung away from Breuil's technique of putting some spirit into tracings (the 'artistic approach'); we now have teams producing 'collaborative copies', or individuals making accurate but often lifeless versions of the Ice Age images (the 'cartographic approach'). Both methods have their merits and their disadvantages.

Deciphering or copying images on a cave-wall is rather like an excavation, except that the 'site' is not destroyed in the process; the pictures are 'artefacts' as well as art and, if superimposed, they even have a stratigraphy.[37]

Moreover, instead of selecting and completing animal figures from the mass of marks, like early archaeologists seeking, keeping and publishing only the 'belles pièces' and ignoring the 'waste flakes', the aim for the last 20 years has been to copy everything. This helps to reduce psychological effects akin to identifying shapes in clouds or ink-blots: faced with a mass of digital flutings or engraved lines, the mind tends to find what it wants to find, in accordance with its preconceptions, and often detects figurative images which are not really there.[38] In addition, one needs to counteract

Fig. 25 *Exhaustive tracing of engraved stone from Gönnersdorf (Germany), and a horse figure 'extracted' from the mass (see also Fig. 93). Magdalenian. The plaquette is 20 cm long, 18 cm wide, and the horse 8.5 cm high. (After Bosinski & Fischer)*

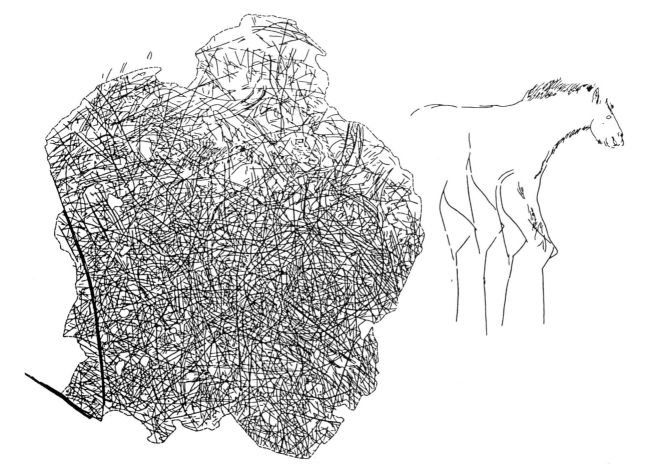

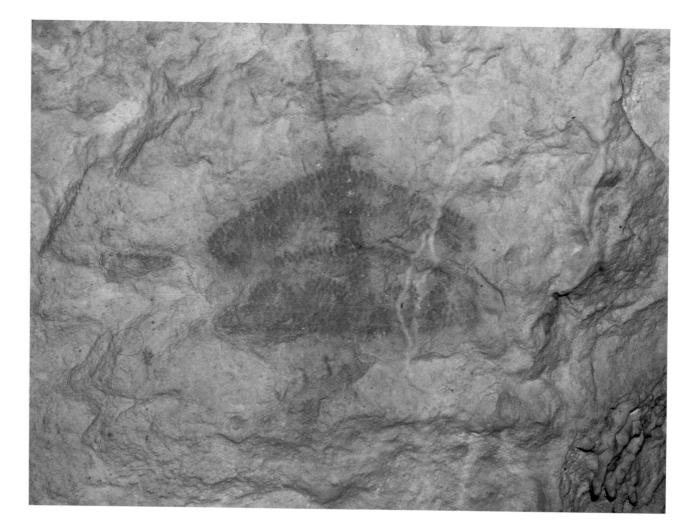

the psychological effect whereby the eye is drawn to the deeper lines (although these may have been of secondary importance) and to lines in concave areas which are generally better preserved than those on convex areas which are more exposed to wear and rubbing.[39]

To eliminate lines we do not understand is an insult to the artist, who did not put them there for nothing; where there are so many lines that it is difficult to 'isolate' anything, however, it is still necessary to 'pull out' any definite figures which exist hidden in the complex mass (this is also far less of a strain on the eyes), though one should still try to publish the mass, leaving the reader free to make a different choice (Fig.25).[40] In his herculean 25-year study of the 1,512 slabs from La Marche with their terrible confusion of engraved lines, Léon Pales isolated and published only those figures which his expert knowledge of human and animal anatomy revealed to his eye: but he estimated that only one line in 1,000 has been deciphered on these stones.[41] Unfortunately, there are very few scholars with similar skills in deciphering and reproducing Palaeolithic engravings.

Breuil's notion of 'traits parasites' has been completely abandoned: Palaeolithic images are not restricted to the figurative but also include non-figurative, abstract and geometric marks, some of them of enormous antiquity, going by the new Australian evidence (see above, p.28).

Just as artefacts are no longer dug up for their own sake, but for what

Fig. 26   *Infra-red photo of painted tectiform at Bernifal (Dordogne). Probably Magdalenian. Width: c 30 cm. (JV)*

Fig. 27   *Drawing of full rock face showing engravings of horses. Montespan (Haute Garonne). Probably Magdalenian. (Drawing by Rivenq)*

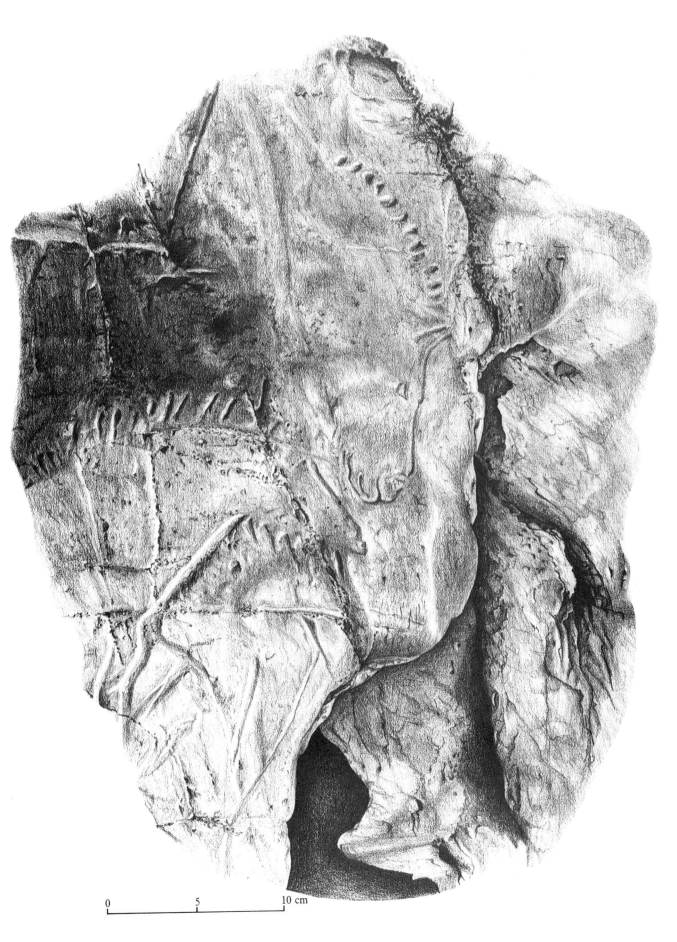

their manufacture, context and associations can tell us, so Palaeolithic images are now studied with the same aims. We shall examine techniques of execution in a later section (p.91), while associations of figures with one another and with differently shaped panels loom large in recent interpretations of cave-art (p.176).

As for context, it is obviously necessary for cave-plans to be as accurate as possible; however, as with tracings of figures, versions have sometimes been published which are so divergent that four plans of La Baume-Latrone, for example, looked like different caves![42] Mistakes have also been made in many cases concerning the part of the cave or the particular wall on which figures occur, and many figures are often missed out altogether.[43] All of this requires checking and published corrections.

We have already seen that copying now includes as much as possible – and it is no longer always restricted to artificial marks, figurative or non-figurative, since it can often be difficult (and occasionally impossible) to tell whether lines are incisions or cracks,[44] or to differentiate between faded spots of paint and blobs of natural colour in the rock.

Photographs, of course, play a major role in studies of context, since a normal lens sees roughly the same area as an eyeball, and a photo shows everything at once: artificial lines, natural marks and fissures, the grain of the rock, and the overall aspect of the panel. In recent publications of some decorated panels, the whole rock-face has been drawn, though few specialists have the ability or the time to do this[45] (Fig.27). A different technique is 'morphometric cartography': ie the making of a detailed contour map, such as that of the panel with the black painted frieze at Pech Merle.[46]

There is also a new emphasis on the physico-chemical interactions of images and wall:[47] close study of the changing conditions in the caves, and of the natural phenomena affecting the images, helps to assess what has happened to them through time, and provides crucial information for conservation.

In short, since the time of Breuil, the task of copying Palaeolithic art has not only acquired many improved techniques, it has also adopted new aims. It is no longer a question of simply accumulating collections of animal figures for publication. All copies, whether tracings or photographs, are seen as tools for further research – as starting points, not as ends in themselves. Tracings are no longer seen as faithful reproductions, but as explanations and interpretations of the images, incorporating an inevitable degree of subjectivity, distortion and choice. Consequently, no copy can ever be definitive, no cave-art can ever be entirely known.

4

How Old is the Art?

How do we know all this art was made in the Ice Age, and during which part of the Upper Palaeolithic? Surprisingly, the methods have changed little since the nineteenth century, and much remains uncertain.

Portable art

Where portable objects are concerned, their position in the stratigraphy of a site, together with the associated bone and stone tools, give a pretty clear idea of the cultural phase involved; of course, the development of radiocarbon dating since the last war has led to fairly accurate measurement of the age of some of these levels – though measurement of the radiocarbon content of a piece of portable art is almost never done because of the destruction involved (in any case, this would date the death of the animal whose bone or antler was used, rather than the artistic activity). Examples include the date of 14,280 years before the present for the layer of La Marche containing the engravings, or that of 11,900 BC for the layer at Enlène which yielded an antler baton with a salmon in 'champlevé', (see below, p.80), or dates of 10,430 and 10,710 BC for the Gönnersdorf habitations containing the engravings.[1]

There are a number of snags. First, many of the best-known pieces of portable art were found in the last century, and they were dug up like potatoes; only a few people like Piette noted the particular layer from which the objects came: early excavators such as Lartet and Christy, or the Marquis de Vibraye, paid comparatively little attention to exact provenance, being carried away by the search for more Palaeolithic 'nuggets'.[2] Consequently, lots of famous specimens can be given no more definite attribution than 'probably Magdalenian'.

Second, a great many objects have no details of provenance at all – either because they come from clandestine digs, or because their sites were atrociously excavated or never published. As we shall see, very few of the 'Venus figurines' of western Europe have any stratigraphic context whatsoever. At Lourdes (Hautes-Pyrénées), over 2,000 cubic metres of deposits were removed from the cave of Les Espélugues (so that it could be turned into a chapel) and scattered over nearby land – the many items of Palaeolithic portable art subsequently discovered in these discarded sediments thus have no stratigraphic provenance.[3]

Third, while a layer may accurately date art-objects such as plaquettes which seem to have been made and discarded quickly, there are many portable objects, notably statuettes, that seem to have been much handled (a point brought out clearly in analyses by Alexander Marshack, see below, p.184) and which may have been carried around and used for generations: consequently, their final resting place may be very far from their place of manufacture, and the layer in which they are found is merely an indication of when they were lost or discarded, not necessarily of when they were made – they may in fact be considerably older.[4]

Finally, there are inevitably occasional disagreements about the precise attribution of the cultural material associated with art-objects: for example, the industry with the engraved pebbles of La Colombière was thought to be an atypical Gravettian, but is now reckoned to be an atypical Magdalenian, which agrees better with its radiocarbon dates around the thirteenth millennium BC.[5]

Much portable art without good stratigraphic context has been assigned to different phases on the basis of style, but this involves subjective judgements and relies on 'type fossils' which often have limited validity. As will be seen (p.85), all the western 'Venus figurines' have unjustifiably been lumped into two groups, the Gravettian and the Magdalenian. It is a tautology to date an object by its resemblance to standard examples, and then to use that resemblance as proof of the date.

Nevertheless, in a few cases, the resemblance is so striking that there is a strong probability that the same period is involved, and at times even the same artist or group of artists; this is shown in examples where both objects being compared are securely dated, such as the spear-throwers of Mas d'Azil and Bédeilhac,[6] or the portable engravings from the grotte des Deux Avens (Ardèche) which are very similar in style and execution (especially of fish) to those of La Vache (Ariège): both have been dated to the mid-eleventh millennium BC.[7]

Chronological schemes tracing the development of portable art have been devised, but tend to be simplistic: one of the earliest was that of Piette, who thought that sculpture in the round came first, to be followed by relief sculptures, 'contours découpés' (p.81), and finally engraving.[8] Unfortunately, he ignored or forgot the fact that engravings, for example, already existed in abundance in his pre-engraving phases! As we shall see, a variety of techniques were already in use before the Upper Palaeolithic, while some (such as the 'contours découpés') seem to characterise certain regions and periods and cannot therefore be taken as fixed stages in the development of portable art as a whole.

Parietal art

Wall-art is extremely difficult to date; as mentioned earlier, attempts are underway to develop means of dating the varnish of patina which forms over rocks and their engravings in desert areas of the world, or, in Australian caves, the layers of carbonate between different phases of engraving on a wall; and since some black drawings at Niaux, Las Monedas and other caves were done with charcoal, rather than the usual manganese dioxide, it may prove possible to date the pigment directly by the radiocarbon method – as with portable art, this would date the death of the tree rather than the artistic activity, but the two events are unlikely to be very dissimilar in age. In all other cases, however, different methods are required.

Proof of Palaeolithic age can take a number of forms: for example, the depiction of animals which are now extinct (such as mammoth), or which were present only during the Ice Age (figures of reindeer in southern France and northern Spain), is a solid argument which, it will be recalled, first convinced the scientific world of the reality of Palaeolithic portable art.

More recently, attempts have been made to date Palaeolithic parietal art by the species depicted: it has been assumed that the pictures reflect the faunal assemblage outside, and that therefore the cave of Las Monedas, which has mostly horses (42%) with some reindeer (13%), represents a cold phase, whereas the neighbouring cave of Las Chimeneas, which has 42% cervids, few horses and no reindeer represents an older, warmer phase[9] – yet both sets of figures seem to be of the same style.

However, whereas Monedas and Chimeneas seem to be simple, homogeneous 'sanctuaries', matters become far more complicated for heterogeneous collections of animal figures, such as the 155 of Castillo which have been assigned to the Aurignacian (27), Gravettian (8), Solutrean (25), Lower Magdalenian (88) and Upper Magdalenian (7); the ten species represented are not particularly indicative of a cold or warm climate; and if the chronological attributions are correct, then there was a massive presence of red deer in the Solutrean, whereas only bison existed in the Upper Magdalenian![10]

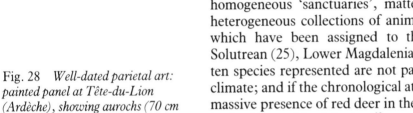

Fig. 28 *Well-dated parietal art: painted panel at Tête-du-Lion (Ardèche), showing aurochs (70 cm long). Solutrean. (JV)*

Clearly, there are many other possible reasons for these differences in depicted species; they should not be taken simply as a tally of what was available outside. As will be seen in a later chapter, species percentages in depictions rarely correspond to those in animal bones; both assemblages are a conscious selection of what was available. While a picture of a reindeer may prove that the animal was present, an absence of reindeer pictures does not prove that the animal was no longer around; and in any case, parietal and portable art of the same period, and even in the same site, tend to depict different species – thus, at the Tuc d'Audoubert, there are close analogies between the two art-forms in style and details of execution, and the two are clearly contemporaneous. The portable material has been dated to 12,400 BC. Yet the bison dominates on the walls, and felines and reindeer are present, together with 100 claviform signs, but there are no fish or anthropomorphs; the portable art, on the other hand, has no claviforms, felines or reindeer, but it does have fish and anthropomorphs, and it is dominated by horses![11] Other examples of differences in the overall content of the two art-forms will be given in the next chapter.

In the past, it was often assumed that a formation of stalagmite or calcite on top of a painting or engraving was proof of great age, but nowadays we are more cautious, since such layers can form very quickly under certain conditions, and may merely prove that the figures are not recent fakes. Very different rates of stalagmite accumulation can occur in the different microclimates of a single cave.[12]

Quite a few caves seem to have been blocked during or just after the Ice Age – Fontanet's entrance, for example, collapsed during the Magdalenian, sometime after 11,860 BC (a date obtained from some charcoal in the cave); and it was occupation deposits of the Palaeolithic which blocked and masked the decorated gallery at La Mouthe – consequently any art located behind blockages of this type has to be of Palaeolithic date.

Similar proof occurs in cases where all or part of the decorated walls themselves are covered by Palaeolithic deposits (datable through their bone and stone tools, and sometimes by the radiocarbon method). The classic example of this type is Pair-non-Pair, whose engravings appeared only when the Gravettian occupation layers were removed; in the same way, the sculptured friezes of Cap Blanc, La Chaire-à-Calvin and Angles-sur-l'Anglin were masked to some extent by Magdalenian deposits; and it will be recalled that the engravings of the Early Man rock-shelter in Queensland disappeared into a layer dating to 13,000 years ago.

There are also a few cases where a fragment of decorated wall has fallen and become stratified in the archaeological layers; at Teyjat (Dordogne), for example, a block of stalagmite with engravings on it was partially covered by late Magdalenian deposits, and a detached portion bearing a bison engraving was found in the lower layer ('Magdalenian V'). We shall see that fallen decorated fragments may constitute a large part of the art from Aurignacian and Gravettian layers.

All the above circumstances, however, merely provide a minimum age for the parietal art. The Pair-non-Pair engravings have usually been attributed to the Gravettian, but they could just as easily be Aurignacian (they were above the level of the Aurignacian occupation layers). The same applies to the Laussel bas-reliefs, which were partly covered by Gravettian layers, but could be older. Similarly, the Isturitz engravings were above the Solutrean level and covered by Magdalenian layers – yet they have often been assigned to the Magdalenian.

A piece of wall may fall to the ground years, centuries or even millennia after it was decorated. Its stratigraphic position records its destruction, not its execution. Only in exceptional cases does the opposite occur: at the abri du Poisson, one of the excavated layers contained thin pieces of limestone which had crumbled from the ceiling; since the sculptured fish itself bore no signs of any such deterioration, it must have been carved after the crumbling (dating to 'Perigordian IV', a Gravettian phase);[13] here, therefore, we have a rare example of a maximum possible age, rather than a minimum.

A different type of maximum age occurs in some high valleys, such as that of Vicdessos (Ariège), where excavation has shown that caves were not occupied before the mid-Magdalenian because of glacial activity; therefore the art in caves in or near that valley, such as Niaux, Fontanet or Les Eglises, cannot be older than that period. But how can one obtain a more precise date?

One way is to find out if and when the decorated cave or shelter was occupied. Some sites have no known Palaeolithic occupation (for example Monedas, Chimeneas, Pindal, etc); in others the deposits were removed, unexcavated, to make visits easier for tourists; and other caves were visited on many occasions in the Upper Palaeolithic (Castillo, Font de Gaume, etc). In some cases, however, there is only one brief period of occupation, and it is therefore likely – though by no means certain – that the artistic activity coincided with this occupation. For instance, the recently discovered decoration of Font-Bargeix (Dordogne) was probably done towards the end of the Magdalenian, since the only known Palaeolithic occupation belongs to 'Magdalenian VI'.[14] The cave of Gabillou had atypical early Magdalenian material at its entrance,[15] and it is thought extremely likely that its very homogeneous collection of parietal figures can be attributed to that occupation, and perhaps even to one person.

Some caves – notably in Cantabria (La Pasiega, Chufín, El Buxú, etc)[16] and the Rhône valley (eg Chabot) – only have Solutrean deposits. At the Tête-du-Lion, the charcoal fragments of the *'foyer d'éclairage'* (see below, p.109) were next to some spots of red ochre on the ground, analysis of which proved them to be of exactly the same composition as the bovid painting on the wall – the radiocarbon date of 19,700 BC therefore dates this cave's decoration to the early Solutrean, like much of the art in this region.[17] Since some caves were decorated but never inhabited, the two phenomena were clearly sometimes separate, so that a single occupation layer, while suggestive, may be much younger or older than the art.

The situation improves if evidence of artistic activity can be found in a site's occupation layers – especially colouring materials in a painted site, as at Tête-du-Lion. This is the case at Altamira, Lascaux and Tito Bustillo, for example; however, one has to be careful how occupation layers are dated: radiocarbon dates from charcoal in Lascaux indicate the early Magdalenian, c 15,000 BC – but, as has been pointed out elsewhere, this is rather like dating a church by analysing the residue from its candles! The study of pollen from occupation layers may also help to assess the date.[18]

A different type of artistic evidence from the occupation levels is portable art; at some sites which have not only parietal art but also well-stratified portable art, one can see clear analogies between the two in technique and style; this method has been in use since the beginning of the century – indeed the first monograph on Altamira devoted an entire chapter to the shared engraving techniques of the two art-forms.[19] In some cases it is quite probable that the same artist was responsible, and at any rate the method can provide a fairly reliable date for the wall-art: at Gargas, there are some resemblances between one or two Gravettian portable engravings and those on the walls. Examples abound in the Magdalenian – the Tuc d'Audoubert has already been mentioned, and in the same way there are close analogies between the portable art of Enlène and, beyond its passage, the parietal figures of Trois-Frères, so that many of the latter can probably be dated to the twelfth millennium BC; engravings on plaquettes at Labastide which are in the same style as its parietal figures[20] come from a layer dated to 12,310 BC.

At Angles-sur-l'Anglin, a 'Magdalenian III' occupation was established on bedrock, its upper surface eventually touching the base of the sculptured frieze, while the 'Final Magdalenian' layer covered the sculptures completely. The artists clearly belonged to the Magdalenian III phase, especially as it featured massive picks, lumps of pigment, ochre crayons, grinding stones and spatulas; in addition, a portrait of a man, engraved on a plaquette from this layer, bears a close resemblance to the big polychrome specimen detached from the site's parietal art,[21] and this phase has been dated to 14,160 years before the present (as mentioned above, the Magdalenian III engraved stones of La Marche, only 30 km away, have been dated to 14,280 before the present, indicating that the two sites were roughly contemporaneous, and there was clearly contact between them. Angles, like La Marche, has limestone plaquettes with animals engraved on them).

The best-known examples of identical portable and parietal figures are the engraved hind heads found on deer shoulder-blades at Altamira and Castillo, and on the walls of both caves. The respective excavators claimed that the Altamira specimens came from a final Solutrean layer and the Castillo ones from an initial Magdalenian, and since then there has been debate about the reliability of these observations, especially as most scholars were keen to include these figures in the Magdalenian.[22] The culprit, as is often the case, was the artificial nomenclature created by prehistorians: it has become clear from more recent excavations that there was continuity between the two phases, and in fact they are probably contemporaneous in Cantabria.[23] It is more sensible to refer to this period by its date (c 13,550 BC at Altamira) than to attach cultural labels to it.

Stratigraphy of figures

Although the Altamira ceiling has sometimes been taken as a single accumulated composition, it actually comprises a series of superpositions: Francisco Jordá has distinguished five separate phases of decoration, beginning with some continuous-line engravings, followed by figures in red flat-wash, then some multiple-trace engravings, some black figures, and finally the famous polychrome paintings.[24] Since the multiple-trace engravings (mostly heads of hinds, and a few other animals) are identical to the portable specimens from the cave, it is clear that the two earlier phases of ceiling-decoration predate 13,550 BC, while the black figures and polychromes are younger (Fig.7).

It was the cave of Marsoulas which first inspired in Cartailhac the idea for dating by superposition – he noted the different styles present and thought he could distinguish at least three layers: black animal figures,

Fig. 29 *Part of the painted panel at Tito Bustillo (Asturias), showing a horse and reindeer, both engraved and painted. The reindeer's muzzle is superimposed on the black horse. The deer measures just over 1 m from muzzle to antler-tip. Magdalenian. (Photo A. Moure Romanillo)*

polychromes, and finally red figures (Fig.8).[25] Subsequently Breuil adopted and extended this approach.

Nevertheless, it is tricky to use superposition as a chronological guide – quite apart from the problem of establishing the order in which layers were applied, which (see below, p.106) can be very difficult. Theoretically, of course, all the layers could have been produced within a very short space of time – a superposition could represent half an hour! – but it is perhaps more likely that they span a few years, at least, and the timespan could be decades, centuries or even millennia.

A problem similar to the Altamira ceiling is posed by the great painted panel of Tito Bustillo, where nine superimposed phases of engraving and painting have been differentiated, culminating in the famous polychrome horses and reindeer. The last five phases are thought to date to the thirteenth millennium BC, on the basis of comparisons with engraved plaquettes from the cave and from the occupation at its entrance, and dates obtained from the layers containing residues of artistic activity at the foot of the panel (stone tools with traces of pigment, etc);[26] but the first four phases are more difficult – the excavators believe that they are probably not much older than the rest, but since phase IV comprises engravings resembling the multiple-trace type, it is possible that it dates to the fourteenth millennium as at Altamira, with the other three phases being older still (Fig.29).

Even more problematical is Lascaux which, in recent years, has been treated as a homogeneous collection of figures, all produced within about 500 years; but Breuil said he could discern 22 different episodes of decoration in the cave, while the abbé Glory saw six in the Hall of the Bulls alone.[27] There is certainly some heterogeneity of style in the cave. The dates of around 15,000 BC for the cave come from charcoal (notably from a layer down the 'well-shaft') and are supported by geometric signs carved on some objects (a lamp, and spearpoints) which are similar to those engraved or painted on the walls, and by the pigments and worn flints in the thin archaeological layer.[28]

There is no doubt that much of Lascaux's art can be attributed to this period; but this does not prove that the whole thing is a coherent entity spanning only a few centuries. It is worth bearing in mind that there is another cluster of dates from the cave in the ninth millennium before the present, and that the remarkable stylistic resemblances between some of its figures (notably the bulls and some deer) and those of some Spanish Levantine art-sites such as Minateda and Cogul, which Breuil noticed immediately and mentioned in print,[29] have yet to be satisfactorily explained (Figs.30-2).

Dating by style

If a decorated site was unoccupied, or simply has no portable art of its own, it becomes necessary to seek stylistic comparison with material from other sites and even other regions. As with undated portable art, one inevitably encounters all the problems of subjectivity, of 'type-fossils', and of over-simplistic schemes of development. All stylistic arguments are based on an assumption that figures which appear similar in style or technique were roughly contemporaneous in their execution.

Suggested sequences of appearance of different forms of representation tend to be highly subjective, and sometimes twist the facts in order to make them fit – for example, Stoliar's scheme, based on only a handful of sites, progresses from the exhibition of part of an actual animal body in the Lower and Middle Palaeolithic, through life-size dummies of various types to sculpture, bas-relief and finally engraving in clay. It requires the Montespan clay statues to be early Aurignacian or even Châtelperronian in date, though there is not the slightest evidence for this view, and the figures are almost certainly Magdalenian like the rest of the cave's art.[30]

Breuil's cycles

Breuil based much of his chronological scheme on the presence or absence of 'twisted perspective', a feature which he considered primitive, and which means that an animal figure in profile still has its horns, antlers, tusks or hoofs facing the front (rather like Mickey Mouse's ears, which always face forward no matter what the position of his body, since this makes animation easier.

It can be argued that this graphic convention is not necessarily primitive, since it enhances the impact and beauty of features like horns and antlers,[31] and can help to suggest depth; indeed, it occurs frequently in the art of later cultures in various parts of the world, up to and including western modern art.[32] Moreover, in the Ice Age figures it is possible that hoofs, at least, were drawn in twisted perspective so that they resembled the animal tracks which must have been of fundamental importance to the hunters and those to whom they had to teach their skills – a side-view of a hoof shows nothing.[33] Indeed, it has been said that if one were to wipe out the bison on the Altamira ceiling and leave only their feet, a professional hunter would at once recognise them as a good representation of a bison's spoor.[34]

For Breuil, however, twisted perspective was a decisive indicator of archaism: he had observed that, at Gargas, the supposedly Gravettian engraved figures had horns seen from the front; the bison on the wall of La Grèze was engraved with horns in twisted perspective, and since this cave had late Gravettian and Solutrean deposits, he thought the bison was relatively early; but in Magdalenian figures, hoofs and horns were drawn in proper perspective. He therefore developed an 'evolution of perspective' which included a 'semi-twisted' stage at Lascaux.[35]

Nevertheless, like Piette's, his scheme was inconsistent, since twisted perspective is also known in the Magdalenian, as in the hoofs of the Altamira bison mentioned above. It is also dominant in the cave of Gabillou which, as we have seen, is probably attributable to the early Magdalenian. The Magdalenian parietal male portrait from Angles-sur-l'Anglin, mentioned above, has a 'full-frontal' eye, as in Egyptian and Cretan art.[36] Breuil himself placed the portable engraving from Laugerie-Basse known as the 'Femme au Renne' firmly in the mid-Magdalenian (its stratigraphic position was uncertain), but the animal profile's hoofs face the front! Yet he claimed that a bovid head with twisted perspective from Isturitz, found in a final Magdalenian layer, had been brought up from a Gravettian level![37] On the other hand, true perspective is sometimes found in early phases, as in some ibex at Gargas, or the deer of Labattut.[38] Twisted perspective is not, therefore, a reliable chronological marker.

Breuil began by proposing a four-stage scheme, but eventually he conceived a development in two cycles, the 'Aurignaco-Perigordian' and the 'Solutreo-Magdalenian', which were largely inspired by the art in the cave of Castillo, and are now of only historical interest, especially as they virtually ignore portable art. With the benefit of hindsight, it seems too neat and symmetrical that Palaeolithic art should have developed as two essentially similar but independent cycles of evolution: each of them progressed from simple to complex forms in engraving, sculpture and painting; each started with 'primitive' or archaic figures, and led on to more complex and detailed images. The second cycle was derived from, and built on the first; overall, he saw a progression from schematic to naturalistic, and finally to degenerate forms.[39]

The 'Aurignaco-Perigordian' cycle featured hand stencils, macaroni finger-markings, simple outline animals with legs omitted, and large red animals in flat-wash (as at Altamira), eventually leading to bichrome and giant figures (as at Lascaux, which Breuil thought Gravettian) with an improvement in perspective from twisted to semi-twisted. He thought that finger markings led to engravings, and thence to bas-reliefs like those of Laussel.

The 'Solutreo-Magdalenian' cycle again began with simple outline animals, then black figures with flat-wash, followed by infill and hatching, and finally the bichromes and polychromes, as at Altamira and Font de Gaume. The sculpture of the Solutrean (such as the carved blocks of Roc de Sers, found in a late Solutrean level) led to the delicate, detailed engravings of the Magdalenian. The cycle ended with the abstract designs of the Azilian. He believed that his system was applicable almost unchanged to all regions with Palaeolithic decorated caves.[40]

Clearly, this system posed problems; we have already seen that a rigid application of the twisted perspective criterion led him to inconsistencies and to errors, such as the wrong dating of Lascaux. Another puzzle was why the bas-reliefs of the Gravettian should be separated in this way from those of the Solutrean, when it was far more likely that they were a continuum.

One solution was proposed by Annette Laming-Emperaire;[41] her approach was somewhat different from Breuil's, since she believed that superpositions were actually purposeful compositions representing a group or a scene. She simplified the evolution of Palaeolithic art into three basic stages: her 'archaic phase' was roughly equivalent to Breuil's first cycle, and included the simple outline animals, for example; her middle phase covered the material where Breuil's two cycles overlapped – ie the bas-relief sculptures, flat-wash paintings, and so forth; and her final phase included the polychromes and other characteristic Magdalenian figures. A division into three 'cycles' was proposed by Francisco Jordá, primarily for Cantabria: they covered the Aurignaco-Gravettian, the Solutrean and lower Magdalenian, and the Magdaleno-Azilian.[42]

Leroi-Gourhan's styles

All these schemes were soon superseded, however, by that of Leroi-Gourhan,[43] which was based primarily on the characteristics of what seemed to be securely dated figures (both portable and parietal) and which still reigns supreme. Like Laming-Emperaire, he believed that superpositions were deliberate compositions, spanning a very short period, and therefore he treated the Altamira ceiling, for example, as a coherent entity – this is probably valid for its polychrome figures but, as mentioned above, there are four earlier phases of decoration beneath them, and it is very risky to assume that they are all contemporaneous, especially as more than one occupation is represented in the archaeological layers of the cave.

Leroi-Gourhan proposed a series of four 'styles', the most recent pair being further subdivided into 'early' and 'late' phases. Unlike Breuil's cycles, they were seen as an unbroken development, with a series of 'pushes' separated by long periods of transition.[44] Following a 'pre-figurative' phase covering some of the Mousterian material mentioned in Chapter 5, Style I comprises material from the Aurignacian and early Gravettian, and notably the carved blocks from those periods, and the fallen fragments of what may have been decorated walls, together with the very few pieces of portable art which he attributed to this phase, such as a few plaquettes from Isturitz. The motifs include deep incisions, figures with stiff contours (the Belcayre herbivore being the only complete specimen), an apparent obsession with vulvas (see below, Chapter 7), an absence of decorated utilitarian objects, and parietal art only in daylight areas of caves and shelters.[45]

Style II corresponds to the rest of the Gravettian and part of the Solutrean; it features good animal profiles, with a sinuous neck/back line, often an elongated head, an oval eye, and twisted perspective, but with the extremities rarely depicted – the bison of La Grèze characterises this phase. Style II includes far more portable art than parietal – partly because so many 'Venus' figurines have been lumped into this phase, and partly because there are so few portable animal figures of the period with which parietal figures could be compared – in the Franco-Cantabrian

Figs. 30/31/32   *Three photomontages of parts of Lascaux (Dordogne). The upper two are from the 'Diverticule Axial', while the third is of part of the 'Hall of the Bulls'. The first includes a falling or jumping aurochs cow (1.7 m long) with a bull's head above, as well as ibex, quadrangular signs, and small horses.*

The second includes deer, aurochs, dots and quadrangular signs, and the three 'Chinese horses'. The second horse is 1.5 m long, the large cow 2.8 m, and the deer on the right 1.5 m.

The third shows the 'unicorn' on the left (1.65 m long), as well as the great aurochs, horses, and some small deer figures. Probably Magdalenian. (JV)

region as a whole, it is reckoned that only about 20% of all portable art predates the mid-Magdalenian! Indeed, in Cantabria, all but one or two pieces are attributable to the late Solutrean or Magdalenian.[46] What little parietal art there is in this phase is still restricted to daylight zones.

Style III, covering the rest of the Solutrean and the early Magdalenian, took as its prototypes the animal figures in caves such as Lascaux and Pech Merle, with their undersized heads and limbs, and semi-twisted perspective dominating: the recently discovered figures in the caves of Domme have been attributed to Style III because of their striking resemblance to depictions at Pech Merle – particularly the volume of the front half of bison-figures, and their elongated bodies.[47] It is in this phase that decoration of the dark parts of caves seems to have begun.

Finally, Style IV includes all the wonders of mid- and late-Magdalenian art, and the decoration of really deep galleries, sometimes marked to their furthest accessible points. The early part is closer in spirit to Style III, with little movement in the figures, which are simply 'suspended in mid-air'; but the more recent phase features more supple animals. In the later part of Style IV, Leroi-Gourhan believed that the decoration of caves gave way to that of plaquettes and similar objects. Contrary to popular belief, the art of the Azilian is not limited to spots on pebbles, but also has naturalistic animal figures in some areas (for example, in portable engravings at Pont d'Ambon, the abri Morin, and the abri Murat).

Leroi-Gourhan's scheme, therefore, is fundamentally like Breuil's in that, with the outlook of modern people, it sees an overall progression from simple, archaic forms to complex, detailed, accurate figures of animals, while signs develop from simple and naturalistic to abstract and stylised forms. It treats Palaeolithic art as an essentially uniform phenomenon. Diversity is played down in favour of standardisation, and the development is greatly oversimplified.[48]

It should be stressed, however, that Leroi-Gourhan, like Breuil before him, was fully aware of the tentative nature of his scheme, and that his Styles had very blurred boundaries. He was able to draw on more securely dated examples than were available to Breuil; but since, as mentioned above, little early parietal art can be dated, he still had to fall back on some of the same criteria as Breuil, such as twisted perspective. Both found it very hard to apply their schemes to painting.

The difference between their two schemes is therefore not so much one of principle as of results – Breuil had the art spread throughout the Upper Palaeolithic (he saw the parietal art of Altamira as dating from the Aurignacian to the Late Magdalenian), but with a very low output in the Solutrean, whereas Leroi-Gourhan compressed the majority of the art into his last two Styles, and especially IV – in other words, their absolute chronologies may differ, but their relative chronologies are similar.

The ladder and the bush

Palaeolithic art did not have a single beginning and a single climax; there must have been many of both, varying from region to region and from period to period. It is self-evident that within those 20,000 – 25,000 years there must have been periods of stagnation, improvement and even regression, with different influences, innovations, experiments and discoveries coming into play. The same is true of the cultures of the Upper Palaeolithic, which come and go during this period, each with its own

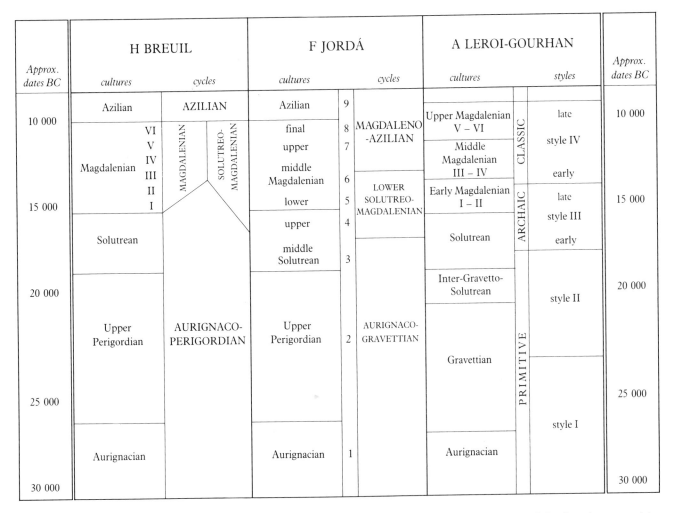

Fig. 33 *Comparison of the dating systems (based on style) of Laming-Emperaire, Jordá Cerdá and Leroi-Gourhan. (After Naber et al)*

specialities (such as Solutrean flintwork, or Magdalenian bonework), constituting a complex interweaving of traditions, with continuities in some aspects of culture and sharp breaks in others.[49]

The development of Palaeolithic art was probably akin to evolution itself: not a straight line or a ladder, but a much more circuitous path – a complex growth like a bush, with parallel shoots and a mass of offshoots; not a slow, gradual change but a 'punctuated equilibrium', with occasional flashes of brilliance. One must never forget that art is produced by individual artists, and the sporadic appearance of genius during this timespan cannot really be fitted into a general scheme. Each period of the Upper Palaeolithic almost certainly saw the coexistence and fluctuating importance of a number of styles and techniques (both realistic and schematic),[50] as well as a wide range of talent and ability (not forgetting the different styles and degrees of skill through which any Palaeolithic Picasso will have passed in a lifetime). It is also naive to assume that all Palaeolithic images are purposeful masterpieces – much of what we call stylisation may simply be compensation for a lack of skill; there must be a certain percentage of meaningless scribble, limited ability, or simply crude attempts by children or beginners, and so we should not assume that everything had complex symbolic meaning.[51] Moreover, there must have been different developments at different times in different regions, and similar styles in two separate regions are not necessarily contemporaneous.

Consequently, not every apparently 'primitive' or 'archaic' figure is necessarily old (Leroi-Gourhan fully admitted this point), and some of the earliest art will probably look quite sophisticated: who, for example, would assign the Vogelherd animals or the Brassempouy head to the Aurignacian if they had not been found in a layer of that period? Leroi-Gourhan preferred to put the Vogelherd animals into his Style II, thus denying their actual provenance.

In some problems, both Breuil and Leroi-Gourhan seem to have been wrong: at Altamira, for example, both of them attributed the polychrome paintings of the ceiling to the late Magdalenian, but excavation revealed only layers of the final Solutrean and the mid-Magdalenian, which were rich in portable engravings and in colouring materials, as we have seen; so it seems likely that all of Altamira's decoration can be attributed to those periods, and that it was blocked during or after the mid-Magdalenian (there is likewise no trace of Aurignacian occupation, contrary to Breuil's theory).[52]

There are many examples where Leroi-Gourhan's scheme proved more satisfactory than Breuil's: for example, at Trois-Frères, Breuil thought several parietal bison in the 'Sanctuary' were Gravettian, but Leroi-Gourhan thought them Magdalenian, and showed they were identical to other bison which Breuil considered Magdalenian! His view has been confirmed by the discovery of a Magdalenian pebble from Enlène with a bison and a horse engraved on it: a broken line next to one figure is very similar to those on seven of the parietal bison.[53]

Other problems remain open, until such time as excavation of the sites in question may help to resolve them: for example, the parietal art of Altxerri seems to be homogeneous, and has been attributed to the late Magdalenian on the basis of Leroi-Gourhan's stylistic criteria, through comparison with portable art from Isturitz, and because the reindeer is depicted; but other scholars have thought it Solutrean and/or lower Magdalenian, and point to supposed resemblances in technique and style between the Altxerri engravings and the 'multiple-trace' engravings of the late Solutrean/early Magdalenian of sites like Altamira and Castillo (and in any case, reindeer bones have been found in Solutrean levels in northern Spain).[54]

It is simply impossible to fit everything into a rigid scheme which minimises the variability of representations in any phase: in Palaeolithic art every rule has exceptions – for example, where all polychrome paintings were automatically thought contemporaneous, it is now clear that those of Altamira are earlier than those of Tito Bustillo. There are no real 'type fossils', and styles are very hard to define and separate. Rather than attempt to define styles, and then fit the images to them, it seems more sensible to take the archaeological and dating evidence at face value, to make an inventory of the different works in each period and their predominant features, and then develop an overall 'evolutionary scheme' (if such a thing is desirable) which incorporates that evidence without distorting it. We must adopt a more flexible system of dating by style, which avoids the problems of rigid systems like those above.

A classic example is provided by the cave of Parpalló, Spain, whose numerous pieces of well-stratified portable art (see above, p.33), spanning many phases of the Upper Palaeolithic, were somewhat neglected by Leroi-Gourhan, perhaps because they display a number of features which contradict his scheme. Work in recent years, particularly in Spain (at Parpalló and other sites), the Rhône valley and the Quercy region, has established that, far from being the artistic desert of Breuil's view, the Solutrean witnessed a tremendous amount of aesthetic activity.[55]

The uncertainties in dating by style alone mean that all known parietal sites are featured on the maps, pp.36-8, with no attempt at dividing them into chronological slices: it is certainly unsatisfactory to compress 25,000 years of wall decoration – two-thirds of art history! – into one map, but this solution is more honest than a subjective division into stylistic periods.

Nevertheless, for the moment, style is the only means we have of dating a great many sites. What, then, is the way forward? Rather than rely blindly on a fixed scheme, or on any particular feature such as twisted perspective, it is wiser to establish assemblages of features: in the same way, archaeological layers are no longer given a cultural attribution because of a single 'type-fossil', but rather on the basis of the whole stone and bone industry present, on the associations, interrelations and presence/absence of different features.

One example is provided by a recent study of a piece of Gravettian portable art, in which the authors drew up a list of definite or extremely probable specimens of portable and parietal art of the period (eg from Parpalló, Isturitz, etc), and then looked for relevant criteria by which they could be recognised.[56] It was found difficult to produce firm criteria for Leroi-Gourhan's styles – as already mentioned, twisted perspective was known in the Magdalenian, while an oval eye is not exclusive to Style II either. However, the piece under study had an association of features (elongated horse head, oval eye, hoof-style) which placed it firmly within that phase.

Despite the variety of the forms which it encompasses, Palaeolithic art does constitute a recognisable episode in art history. We have seen that, for a number of reasons, almost all of it can safely be attributed to the 20 or 25 millennia of the Upper Palaeolithic, and most of it to the last ten. More precise attributions are difficult at present and, where possible, one should therefore try to restrict discussion to examples of portable and parietal art which are securely dated, and thus hope to sidestep accusations of subjectivity and tautology!

5

Forms and Techniques

Before embarking on a survey of those forms and techniques of Palaeolithic 'art' which have survived, it is worth remembering that they probably represent only the tip of the iceberg: although it cannot be proved (unless one day a waterlogged, desiccated or frozen site of the last Ice Age is discovered, in which everything is preserved), it is virtually certain that a great deal of artistic activity involved perishable materials which are gone for ever: work in wood, bark, fibres, feathers, or hides (as mentioned earlier, some decorated wooden rods are known from late Middle Stone Age levels of Border Cave, South Africa, dating to c 37,000-50,000 years ago[1]); figures made in mud, sand or snow; and, of course, body painting which, through finds of red ochre in living sites and in graves, is thought to have very remote origins.

Music

In addition, dance and song leave no traces at all, and such things as reed-pipes, wooden instruments, and stretched-skin drums will have disintegrated; however, a few musical instruments have survived from the Upper Palaeolithic – there are about 30 'flutes', spanning the Aurignacian and Gravettian (18), the Solutrean (3) and the Magdalenian; a handful come from Hungary, Yugoslavia, Austria and the USSR, but most are from France, with 14 from the supersite of Isturitz alone.[2] The majority are broken; the French ones are made of hollow bird-bones, while the eastern specimens are of reindeer or bear-bone; they have between three and seven finger-holes along their length, and are played like penny-whistles rather than true flutes. Experiments with a replica by a modern musicologist have revealed that, once a whistle-head is attached to direct the air-flow, one can produce strong, clear notes of piccolo-type, on a five-tone scale.[3] It should be noted that a possible flute, comprising a fragment of a very small hollowed-out bone into which someone had begun to make two holes, came from a pre-Aurignacian layer of the Haua Fteah cave, North Africa, dating to c 45,000 years ago.[4]

A few shaped, polished and engraved bird-bone tubes have been found which have no holes, and have been interpreted as trumpet-like 'lures' for imitating the call of a hind in the rutting season – one fine example from the Magdalenian site of Saint-Marcel even has a series of what look like cervid ears engraved on it![5] Many perforated reindeer phalanges have been interpreted as whistles in the past, though often the hole was made by carnivore teeth or other natural breakage; those which were intentionally made do produce a shrill, powerful note. A few definite whistles in bird bone are also known, such as the Magdalenian specimens from le Roc de Marcamps, Gironde.[6]

A number of oval objects of bone or ivory, with a hole at one end, have been interpreted as 'bull-roarers' ('rhombes', or 'bramaderas'), a type of instrument which makes a loud humming noise when whirled round on a string: a particularly fine example made of reindeer antler is that from the cave of La Roche Lalinde (Dordogne) (Fig.34). The well-known parietal engraving from Trois Frères of a 'sorcerer' with a bison-head (Fig.104) has often been interpreted as playing a musical bow, but this seems an extremely tenuous idea, and the enigmatic marks in front of its mouth could be any one of a number of things (see later, p.152).

As for percussion, a number of mammoth bones, painted with red ochre, from the site of Mezin, near Kiev, dating to c 20,000 BC, are thought to be musical instruments – a hip-bone xylophone (osteophone?), skull and shoulder-blade drums, and jawbone rattles – and have been played by Soviet archaeologists, who even cut a record of their jam-session.[7]

Finally, there are possible lithophones in a number of caves: 'draperies' of folded calcite formations often resound when struck with a hard object (wooden sticks seem to produce the clearest and most resonant notes), and this seems to have been noticed by Palaeolithic people, since some of the lithophones are somewhat battered, and are decorated with painted lines and dots.[8] Apart from Nerja in Spain and Escoural in Portugal, all known examples are in the Lot region of France (Cougnac, Pech Merle, Les Fieux, Roucadour); moreover, most of them are in or near large chambers which could have held a sizeable audience.

Red skins or tanned hides?

Virtually all peoples around the world paint their bodies on certain occasions, and we have no reason to doubt that the same was true during the Palaeolithic – indeed, this was probably one of the very first forms of aesthetic expression. Unfortunately, owing to the decomposition of bodies, we have to infer it from other evidence. Lumps of natural pigments are known from archaeological sites of very remote periods: for example, one piece of red mineral, with vertical striations resulting from use, was found in the Acheulian (c 250,000 BC) rock-shelter of Bečov, Czechoslovakia, which had been occupied by *Homo erectus*. The even earlier site of Terra Amata (c 350,000 BC) at Nice, France, produced 75 bits of pigment ranging in colour from yellow to brown, red and purple; most of them have traces of artificial abrasion and were clearly introduced to the site by the occupants, since they do not occur naturally in the vicinity.[9]

During the period of Neanderthal people (c 100,000–35,000 BC), such pigments become increasingly frequent, not only in occupation deposits but also in burials, which now occur for the first time. In France, for example, the cave of Pech de l'Azé yielded 103 blocks of manganese dioxide (black/blue), plus 3 of iron oxide (red); 67 of them were rounded or polished into a 'crayon' shape, as if they had been used on some soft surface.[10]

A Neanderthal skeleton at Le Moustier was sprinkled with red powder; red pigment was also found around the head of the famous skeleton of La Chapelle aux Saints, near two skeletons at Qafze, Israel, and with many others. In addition, there is evidence for actual mining of haematite (iron oxide) in Southern Africa from c 45,000–50,000 BC onwards (it will be recalled that pigment was in use in Zimbabwe at least 125,000 years ago), and in Hungary from 30,000 BC.[11]

It was in the Upper Palaeolithic period that pigments became really abundant, being transported in tens of kilos; in some French occupation sites, it is not rare to find habitation floors impregnated with red to a depth of 20 cm. Well over 100 sites with pigment are known, as well as at least 25 burials. As we shall see, some of these pigments can be linked to the decoration of cave-walls; but what about open-air sites or unpainted caves? Can we assume that the presence of pigments here necessarily indicates body-painting?

Unfortunately, matters are not so straightforward, for the simple reason that mineral pigments of this type have a number of properties. Ethnographic studies around the world show that ochre is often used in the treatment of animal skins, because it preserves organic tissues, protecting them from putrefaction and from vermin such as maggots. It is probably this kind of function which explains the impregnated soil in some habitation sites and the traces of red mineral on many stone tools such as scrapers. Similarly, red pigment may have been applied to corpses not so much out of pious beliefs about life-blood, as is commonly assumed, or in order to restore an illusion of health and life to dead cheeks, but rather to neutralise odours and help to preserve the body.[12]

Even if, as most prehistorians believe, the people of the Old Stone Age did indeed paint their bodies, the practice may have been purely functional in some cases, rather than aesthetic: ochre is very effective in cauterising and cleaning injuries, and is still used in parts of Africa to dry bleeding wounds – in fact, until the end of the last century, it was still used by country doctors in parts of Europe as an antiseptic in the treatment of purulent wounds. Another function which may have been important during the last Ice Age is that of protection against the elements and insects. Peoples such as Polynesians, Melanesians and Hottentots used red pigments to maintain bodily warmth and ward off the effects of cold and rain; among some North American Indian tribes, a mixture of red ochre and fat was often applied to the cheeks of women and children as a hygienic measure to protect their skins against the sun and dry winds; while in other parts of the world red paint has been used as a protection against mosquitoes, flies and other disease-carriers.

The more aesthetic uses of pigment on the body may well have arisen from practices such as these: for instance, the medicinal properties of ochre may have led to the painting of the dead or dying, as was commonly done by certain tribes in the New World. In Palaeolithic cases, it is often difficult to tell whether a body had its flesh painted or merely its bones. If the whole body was painted at death, or just before, the lumps of pigment placed with the corpse may represent supplies of body-paint for the afterworld.

Fig. 34 *Engraved 'bull-roarer' with geometric/linear motifs and covered with red ochre, from La Roche at Lalinde (Dordogne). Magdalenian. Length: 18 cm, width: 4 cm. (JV)*

But what of the living? As already mentioned, most authorities agree that Upper Palaeolithic people must have painted their bodies, but the evidence is very limited. The use-wear on lumps of pigment is often shiny, indicating that they were applied to soft surfaces, which could be human skin, but also animal hide; the traces of use on a rough surface may simply indicate removal of powder to be used in a liquid paste. Some French caves have yielded bone tubes or hollow bird-bones, often engraved, containing powdered pigment, and not all of this material can be linked to cave-art; at the Mas d'Azil, excavators found a flat cake of red ochre, pitted with holes, and associated with sharp bone needles, which they interpreted as evidence for tattooing in the Upper Palaeolithic.[13] Finally, as will be seen below, some human figurines of the period were originally painted red – but the same is true of bas-reliefs of horses and fish!

Techniques in portable art

These will be reviewed in order of apparent complexity, but this should not be taken as a chronological progression: as we have seen in the preceding chapter, the phenomenon of Palaeolithic art is a complex web of forms and styles rather than a simple linear development.

Portable art is usually divided into what seem to be utilitarian and non-utilitarian objects (ie decorated tools versus art objects or ornaments), and their decoration is classed as figurative or non-figurative. A very wide variety of materials and forms was employed.

The study of the techniques used in the period rests on two main types of evidence: first, the traces left on the objects or images by the tools, together with precise observation of their technological characteristics (in a very few cases, production debris or what may be the original tools have survived in close proximity to the images); second, experiments with similar materials and tools have been carried out, followed by comparison between the modern results and the originals.

Slightly modified natural objects

Much of what is called 'parure' (jewellery) belongs in this category – ie fossils, teeth, shells or bones which have been incised, sawn or perforated. Such techniques are by no means restricted to the Upper Palaeolithic: a growing number of specimens are known from the preceding (Mousterian) period, and can therefore be attributed to Neanderthalers: two bones (a wolf foot-bone and a swan vertebra), with holes bored through the top, from Bocksteinschmiede (Germany), dating to c 110,000 years ago; a carved and polished segment of mammoth molar, and a fossil nummulite with a line engraved across it (making a cross with a natural crack), from Tata (Hungary), dating to c 100,000 years ago; a bone-fragment from Pech de l'Azé (Dordogne), with a hole carved in it; a reindeer phalange with a hole bored through its top, and a fox canine with an abandoned attempt at perforation, from La Quina (Charente).[14] As will be seen below, other forms of 'aesthetic expression' are also known from the Mousterian.

The earliest phase of the French Upper Palaeolithic, known as the Châtelperronian (c 35,000 BC), has yielded a few more examples: the best known are those from the cave of Arcy-sur-Cure (Yonne). These levels at the site also contained a Neanderthal tooth; in view of the discovery of a Neanderthal skeleton in a Châtelperronian layer at St Césaire (Charente

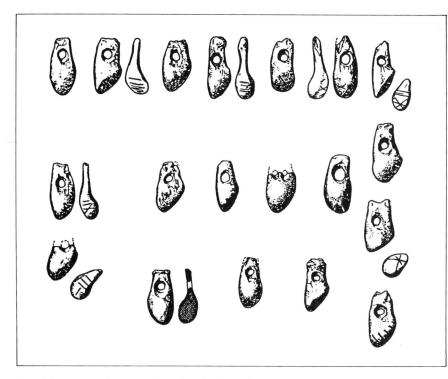

Fig. 35 *Perforated and engraved deer canines from the Magdalenian burial at St Germain-la-Rivière (Gironde). (After Laurent)*

Maritime),[15] it is more than likely that the earliest Arcy ornaments can be attributed to Neanderthal craftsmanship. They include wolf and fox canines made into pendants by incising a groove around the top, at least one sawn reindeer incisor, a bone-fragment with a wide carved hole, a sea fossil with a hole bored through its centre, and a fossil shell with a groove cut around the top.[16]

The next layer at Arcy, representing the Aurignacian (c 30,000 BC), has material which features the same techniques, clearly drawing on what had been developing for millennia: perforated fossils, a bone pendant with a wide carved hole, and so forth. Even older Aurignacian sites, such as the cave of Bacho Kiro in Bulgaria (over 41,000 BC), contain perforated animal-tooth pendants.[17]

The Aurignacian therefore has no sudden appearance of this kind of material, but there seems to be a marked increase in its abundance, perhaps linked to advances in lithic technology which improved or facilitated the working of decorative objects. The three main classes of '*parure*' are beads (of ivory, bone, stone, fossil wood, etc), animal teeth, and shells.

The animal teeth, perforated through the root, are mostly bovine incisors and the canines of fox, stag, wolf, bear or lion – fox teeth are often the most abundant in the Aurignacian and Gravettian, especially in central and eastern Europe.[18] For example, the old man buried at Sungir, near Moscow, about 24,000 years ago, had two dozen perforated fox canines sewn on the back of his cap,[19] and more were found with the two children buried at the site. Over 50 perforated fox canines were found in Kostenki XVII (USSR) and are dated to 32,000 years ago,[20] while 150 covered the head of the child buried beside Kostenki XV. In western Europe, on the other hand, and particularly in the Magdalenian, stag canines were the favoured decoration – the best known are the 70 perforated specimens found round the neck or chest of the woman buried at St Germain-la-Rivière (Gironde), 20 of which are engraved with crosses or parallel lines.[21]

The popularity of canine teeth continued to the end of the Upper Palaeolithic, but in the Magdalenian there was an increase in the practice of sawing reindeer incisors, which (as mentioned above) is already present in the Châtelperronian period at Arcy: over 50 have been found at La Madeleine and Gönnersdorf, and over 200 at Petersfels (Germany). Their occasional discovery in rows shows that, as among some northern peoples in historical times, their roots were sawn through and they were then cut from the mouth as a group, still held inside a strip of gum which was handy for hanging them as a string of eight 'pearls'.[22] Occasionally one encounters the teeth of other species used as pendants: eight perforated human teeth are known from sites in France (one from the Aurignacian site of La Combe, the rest from Magdalenian sites such as Bédeilhac), and one from the Gravettian of Dolní Věstonice (Czechoslovakia);[23] there are pierced seal-teeth from the Magdalenian of Isturitz, and many sites have imitations of canines (especially stag canines) made out of ivory, stone or bone.[24]

As for shells, only a few species were selected: primarily small, globular gastropods (such as *Littorina* and *Cypraea*) which could easily be sewn to clothing; long forms (such as *Dentalia* or *Turritella*) which could be easily strung; and a few scallops (*Cardium*, *Pectunculus*).[25] Many of these species are inedible, and their function was clearly decorative rather than nutritional: most were perforated with a pointed tool. They are often found in considerable quantity, even in early sites – there were 300 in the Aurignacian Cro-Magnon burial alone (and hundreds more in other burials), while living-sites such as Isturitz (Pyrénées-Atlantiques) or the abri Blanchard (Dordogne) contained hundreds of periwinkles. Fossil shells were also utilised and sometimes came from great distances – for example, those at Mezin (USSR), a site dating to 21,000 years ago, came from a distance of at least 600 km.[26] The shells of land molluscs were rarely used, no doubt because they are thinner and more fragile.

Research shows that Dordogne sites (such as Laugerie-Basse, abri Pataud, abri Castanet) generally yield a high proportion of species from the Atlantic, particularly those which are common along the coast of Charente, but shells from the Mediterranean are also clearly represented. One finds the same ratio at the Atlantic end of the Pyrenees; but sites in the Central Pyrenees, such as Lespugue, at a distance of 200 km from either coast, contain a more even ratio, while further to the east, in Ariège, the proximity of the Mediterranean is reflected quantitatively in the shell collections,[27] although even here Atlantic shells dominate slightly – this is no doubt because all rivers in the French Pyrenees (apart from those at the eastern extremity), like those of Dordogne, flow out to the Atlantic, and this must have determined the movement of people and materials to a considerable extent.

It is theoretically possible that all the shells came inland in an exchange network involving 'maritime peoples' for whom we have no evidence whatsoever thanks to the drowning of the coastlines of the last Ice Age through the rise in sea-level since that time. Certainly a great deal of exchange went on: for example, Mediterranean shells have been found in the German site of Gönnersdorf, 1,000 km away, and it is unlikely (though possible) that the site's occupants travelled to that coast for them.

It has been argued[28] that the great number of shells in the sites of Blanchard, Castanet and La Souquette means that this clutch of Aurignacian sites represents a market centre for exotic materials. Equally, however, it could simply mean that these sites were centres of production, or the habitations of those who specialised in working these materials: there is ample evidence throughout the Upper Palaeolithic for craft specialisation, and for repeated contact with the coasts which involved not merely vague 'exchange networks' but also probably the seasonal movements of people, following herds, and dispersing or coming together in certain places at different times of year.[29]

In the past it was often assumed that the shells served as pendants and ear-rings or, in groups, as necklaces or bracelets. However, as with beads it is evident from finds such as the burials at Grimaldi on the Mediterranean coast, or those of Sungir, that many were attached to caps and clothing. At Sungir, for example, the three burials had only a handful of perforated shells, but about 3,500 beads of mammoth ivory each, arranged in rows across the forehead and temples, across the body, down the arms and legs and around the ankles. Rather than being sewn on to garments one at a time, it is far more likely that the beads were strung on lengths of sinew, which were then attached to the clothing.

It has been estimated that a Sungir bead would have required about 45 minutes for its manufacture (cutting the tusk, drilling the hole, etc), which means that each body had 2,625 hours of 'beadwork' buried with it;[30] and the very standardised and uniform appearance of these objects suggests that they were produced by only a few people. In western Europe, far more ivory beads are known from living sites than from burials; only the context provided by a burial can indicate an ornament's true function, but it is likely that most of the finds from occupation sites were also attached to clothing. Some idea of the production sequence involved in beads can be gained from the Aurignacian material at Blanchard, which includes pieces at different stages of manufacture: small prepared rods of ivory were divided into sections, separated into pairs, and then worked into a dumb-bell form and perforated, before the final shaping (most are round or basket-shaped).[31]

Other types of bead include fish vertebrae, which were sometimes strung together as necklaces, as in an example from a burial at Barma Grande.[32] It is worth noting that ornamentation of this kind is by no means restricted to Europe: for example, it will be recalled (see above, p.29) that perforated bone beads are also known from the late Pleistocene of Australia (Devil's Lair), while the Upper Cave at Zhoukoudian, China, dating to 18,000 years ago, yielded over 120 decorative items such as pierced vertebrae of carp and other fish, perforated seashells, animal teeth, small pebbles, and engraved/polished bone tubes cut from the leg-bones of birds. Many of the perforations are coloured red, suggesting that whatever thread they were strung on was dyed.[33]

As with any other category of portable art, there is a marked differentiation in the distribution of ornaments: many sites in Europe (including some burials) have none or a few, and others have hundreds. As mentioned above, this probably reflects the presence of specialised craftsmen, as well as the varying functions of different sites (including clothing manufacture?), and perhaps even, where rich burials are concerned, some form of incipient hierarchy.

Engraved or painted stones
It will be recalled that the engraving of a bear on a pebble from the cave of Massat was one of the first pieces of Palaeolithic portable art to be

Fig. 36 *Engraved plaquette from Le Puy de Lacan (Corrèze), showing a duck-like bird, a fine bison head above it, and, to the right, the hindquarters of another bison. Total width: 20 cm. Magdalenian. (JV)*

authenticated; and that a few sites have hundreds of incised slabs – mostly of sandstone, limestone, slate or stalagmite. Few examples of portable engraving are known from the early Upper Palaeolithic or from eastern Europe (one of the earliest is the geometric motif on a plaquette from the Châtelperronian Grotte du Loup, Corrèze,[34] and a late Mousterian limestone slab from the cave of Tsonskaïa in the Caucasus has a cross engraved on it[35]), and in fact this particular type of art characterises the Magdalenian of western Europe.

The incisions on stone are sometimes deep and clear, but in many cases they are so fine that they are almost invisible – this is why so many engraved pieces are found discarded in the spoilheaps of earlier, less alert excavators, as at Enlène. Only under a strong light coming in from the side can one see the lines at all; indeed, this kind of fine engraving can almost be classed as a drawing rather than an incision!

Alexander Marshack has pioneered the 'technological reading' of Palaeolithic images; his studies of engraved stone, bone and antler under the microscope have enabled him to follow the mechanics, micromorphology and 'ballistic trace' of each incision – its point of impact and subsequent path – and to identify marks made in different ways: from left to right or right to left, as arcs, jabs, etc. He has also claimed that he can detect changes of tool and of hand: for example, he believes that at least seven different points were responsible for a fish with 'arrows' on a stone from Labastide (Hautes Pyrénées), and that four or five different points were responsible for renewing the horns on the rhinoceros of La Colombière.[36]

However, other scholars have argued that, at least where small stone plaques are concerned, a single burin can produce a wide variety of traces on them, depending on the part of the tool used, its position and angle, and the strength of the hand – for example, the section of the incision changes when a straight line is continued as a curve.[37] In any case, a tool may have been resharpened in the course of use. Experiments with slates of the type found at Gönnersdorf and Saut-du-Perron suggest that any changes in the incised marks are due to differences in tool-pressure, not to different tools.[38]

The new technique of placing varnish replicas of engraved surfaces under the scanning electron microscope (see above, p.45) is now being used in an attempt further to elucidate this problem; criteria are being sought which would identify the use of the same tool in different kinds of incisions – for example, it has been found that every tool leaves distinctive secondary striations alongside the main incision whenever parts other than the point have momentary contact with the stone.[39]

A different question concerning 'plaquettes' (defined as slabs of stone with parallel faces, under 20 cm across and 4 cm thick) is that of their usage. Scholars such as Henri Bégouën, who believed in hunting magic (see p.151), were inclined to see evidence of ritual in everything. Since many of the engraved plaquettes from Enlène were broken, and fragments of the same specimen might be found metres apart, he concluded that the breakage and dispersal were purposeful; since most of them had been burned, and appeared to lie with the engraved face downward, this too formed part of the ritual.[40]

0   5   10   15 cm

Fig. 37 *Tracing of an aurochs and deer on a plaquette from Trou de Chaleux, Belgium. Magdalenian. Total width: 80 cm. (After Lejeune)*

It is not always the case that most lie with the engraving face-down (though this seems to be true for sites such as La Ferrassie and the abri Durif at Enval[41]); the fineness of many incisions makes it impossible to see them within the cave, and if plaquettes are collected from a cave-floor, as at Labastide, it is hard to remember which way up they were – in any case, some have engravings on both sides: at Enlène, about 16% are engraved on both sides; and, of 50 which are not but whose position *in situ* was noted, 27 had the engraving face-up and 23 face-down![42]

Moreover, in some caves, the upper face of a plaquette lying on the surface becomes coated with a deposit of clay or calcite and needs to be washed or treated with acid before the engravings underneath become visible.[43] As for dispersal, it is true that pieces have often been well scattered, and in some cases missing fragments have never been recovered; this may sometimes denote purposeful breakage, but people and animals trampling around on cave- or habitation-floors can also have drastic effects on the material lying there.

The question of burning is more interesting. There is a definite link between plaquettes and fire in some caves: many have marks of burning and charcoal, some (as at Labastide) have been found in hearths, while others have even been used in their construction, as at Mas d'Azil.[44] Breuil and Bégouën adopted the view that they had served as crude lamps, and this may be true in some cases. However, a theory has also been put forward that they represent a kind of heating device – sandstone has thermal qualities and a resistance to tension which make it suitable for a function of this kind. In France, old peasants in some areas still use heated sandstone plaquettes, wrapped in cloth, as bedwarmers![45] It may therefore be thermal tension which shattered many stones, not some arcane ritual, although a few (eg at Labastide and Enlène) do bear marks of blows suggesting deliberate breakage.

It is difficult to assess how important these engraved stones were. In sites such as Enlène, Gönnersdorf, Labastide and Enval there are many of them (over 1,000, 500, 50 and 27 are known respectively), but there are hundreds more without engravings in these sites. Enlène has tens of thousands of plaquettes, brought in from a source some 200 m from the cave; excavation has shown that an area of over five square metres of cave-floor was paved with them, no doubt a measure against humidity, and this paving includes engraved specimens;[46] a similar area of cave-floor at Tito Bustillo (Asturias) had 83 plaquettes, 25 of which bore engravings ranging from animal figures to simple incisions.[47] At the late Magdalenian open-site of Roc-la-Tour I (northernmost France), hundreds of schist plaquettes were brought in as paving, but only about 10% have engravings.[48] Similarly, at Gönnersdorf, several tons of schist plaquettes were brought to the open-air site as elements of construction and as foundations for structures; only 5–10% of them were engraved, and these seem to be distributed at random among the others.[49] This suggests that the engravings lost all value once the ritual had been performed and they had been broken and dispersed; or simply that the engravings never had any ritual significance, and were done simply to pass the time, for practice, for storytelling, or perhaps even to personalise one's private bedwarmer! Future finds may help to clarify the situation further.

Occasionally, engraved specimens are stuck vertically in the floor – for instance, around hearths at La Marche; and a few painted stones are known from Enlène, Labastide, Parpalló, etc. Some are clear figures, while on others simple staining by ochre in the soil may be involved: at Tito Bustillo, it was thought at first that some plaquettes had paint on, but careful study revealed that this was all contamination by pigments in the soil;[50] the same phenomenon occurred on the engraved plaquettes of the abri Durif at Enval, where red sand had coloured the side facing downwards, but at La Marche it is possible that some red traces on slabs are paint, while others seem natural.[51]

The best-known painted stones are those which characterise the very end of the Palaeolithic in western Europe – the small 'Azilian pebbles', first identified in the 1880s at the Mas d'Azil by Piette (see p.23). They have been found in 28 sites in France, 5 in Spain, 3 in Italy and 1 in Switzerland; but, of the nearly 2,000 known, over 1,400 are from the Mas d'Azil. Their motifs, usually in reddish ochre, are simple (mostly dots and lines), and seem to have been applied with the finger, less often with a fine brush; but, as will be seen later (p.182), a recent study has produced fascinating results from an analysis of the numbers and combinations of these marks.[52]

It will be recalled (see above, p.27) that painted animal figures are also known on stones from the late Pleistocene Apollo 11 Cave in SW Africa, while engraved pebbles have been found at Kamikuroiwa, Japan (p.28), and non-figurative engraved specimens are quite common in the Azilian of western Europe.[53]

Where engraved stones are concerned, it should be noted that, although it is the fine figurative examples which tend to get published, there are far more which are indecipherable, either because they are tiny fragments or because they are non-figurative (at Roc-la-Tour I, about 600 small engraved fragments have been recovered, but there are only 16 'readable' figures so far[54]). Some have a confused mass of superimposed lines (as on the slabs of La Marche, or the pebbles of La Colombière), but experiments with Gönnersdorf slabs show that a fresh engraving is very visible, due to the presence of white powder in the incisions. When this is washed off, the effect is like wiping chalk off a slate, and a new engraving can be made (the incisions can be made quickly and easily, without effort[55]) – this suggests that some of these engravings had significance for only a very brief time, and the 'associations' of superimposed animals on a given surface are not necessarily meaningful; on the other hand, there were plenty of stones available, and each figure could easily have had one to itself if desired (as is generally the case at Enlène, for example), so the superimpositions may indeed have had some significance.

Painted bone, and engraving on bone and antler
Many of the above comments also apply to incisions on flat pieces of bone; but experiments show that, unlike stone, fresh bone is hard to engrave: the tool tends to skid when it cuts bone fibres, and extremely sharp tools are required. Moreover, it is necessary to pass the tool backwards and forwards, to widen the incision. One of the chief difficulties is that the initial marks can barely be seen, though it has been found that covering the bone with ochre beforehand makes incisions readily visible[56] – this may explain why some Palaeolithic specimens, such as the bone-fragment from Enlène with a grasshopper engraved on it, were covered in ochre when found;[57] decoration of bones with pigment also survives occasionally – it will be recalled that the mammoth-bone 'musical instruments' from Mezin were painted with geometric motifs, chevrons, and undulations.

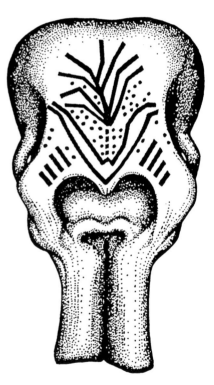

Fig. 38 *Reconstruction of the mammoth skull painted with red ochre, from Mezhirich (USSR). Maximum width: 60 cm. (After Pidoplichko)*

Fig. 39 *Reindeer shoulder-blade from the rock-shelter of Duruthy (Landes) with a reindeer engraved on it. Magdalenian. The animal measures 9.5 cm from muzzle to tail. (PGB)*

Fig. 40 *Detail of a perforated antler baton from Duruthy (Landes) which bears a depiction in 'champlevé' of two ibex, perhaps fleeing a predator at left. Magdalenian. The photo shows one of the ibex (c 4 cm long). (PGB)*

Fig. 41 *Engraved and carved depiction of a bison calf, only a few cm in length, from Le Grand-Pastou (Landes). Magdalenian. (PGB)*

Similarly, a mammoth skull from dwelling No 1 at Mezhirich, Ukraine (c 12,000 BC), is decorated with zigzags and dots of red ochre; engraved lines on bone and ivory in eastern Europe are often highlighted with a filling of black, manganese paste, whereas red ochre tended to be used for this purpose in western Europe.[58]

Shoulder-blades are the bone equivalent of plaquettes, having a smooth, large surface, and it is therefore surprising that they were not engraved in great numbers; nevertheless, decorated specimens are known from some sites, and the Mas d'Azil and Castillo each have over 30, together with undecorated ones; their engravings are both figurative and 'abstract'. Those from the Mas d'Azil, at least, were clustered together in a small area very poor in other finds.[59] At the abri Morin (Dordogne), 17 fragments with figurative engravings have been found, and 29 with non-figurative marks.[60]

Although fibrous and relatively soft, such bones are by no means easy to decorate: Jean Bouyssonie, the French prehistorian, found that it took considerable muscular strength to engrave a fresh horse shoulder-blade, and the flint point often broke.[61] In some cases, the tool does not incise the bone, but compresses its surface into a furrow.

Experiments also indicate that a burin is not the only tool which can engrave bone, although it is the one which is always mentioned in this connection. In fact, a wide range of stone tools are equally effective – awls, pointed backed bladelets, and even the edge of a broken blade are just as good; it is the sharpness which counts, not the precise form. A copy of a small bison engraving on bone from La Madeleine, using different kinds of incisions and tools (which displayed no traces of wear afterwards), took four hours; but a second attempt halved that time, showing that a practised Palaeolithic craftsman would doubtless have taken little time to produce this kind of image – similar results were obtained by Leguay, the pioneer of this type of work, who, over a century ago, tried engraving on bone in the prehistoric way using original stone tools from Palaeolithic sites, and found that it could be done quickly with a little practice.[62] However, we still have much to learn about the engraving of bone and antler, such as whether specimens of different kinds and ages vary in their 'incisability', or how special processes such as soaking may have affected the work.

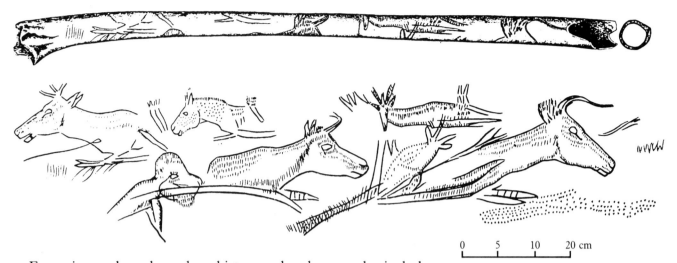

Fig. 42   *Tracing of figures (deer, horse, isard, ibex, aurochs and human) engraved around the gannet bone from Torre (Guipúzcoa). Magdalenian. Length: c 16 cm. (After Barandiarán)*

Engraving on bone has a long history and early examples include: a fragment of bovine rib from an Acheulian layer at Pech de l'Azé, dating to c 300,000 BC, bears an intentional engraving comprising a series of connected double arcs, while several bones from Bilzingsleben (East Germany), of about the same date, have geometric designs engraved on them which are clearly purposeful, and nothing to do with cutting or working;[63] a Mousterian fragment from the shelter of Riparo Tagliente, Italy, also has a double arc incised on it;[64] La Quina, which, as mentioned above, had two perforated pieces in its Mousterian, also yielded a bovid shoulder-blade engraved with very fine, long parallel lines, while a Neanderthal burial at La Ferrassie (Dordogne) contained a small bone with a series of fine, intentionally incised marks which reminded the excavator of the notched bones of the Aurignacian – similar fragments with regular notches are known from the Mousterian of other European sites such as Cueva Morín and several sites in Charente;[65] the Bulgarian cave of Bacho Kiro has a Mousterian bone-fragment with a zigzag motif engraved on it;[66] while a Mousterian layer at Molodova-I, USSR, more than 40,000 years old, has a mammoth shoulder-blade decorated with little pits, patches of colour, and notches that form complex patterns including cruciform and rectangular figures in which some Soviet scholars have seen the outline of an animal.[67] It will be recalled (p.27) that a baboon fibula with 29 parallel engraved notches has been found at Border Cave in southern Africa, and dates to 35,000–37,500 years ago. When faced with examples of this kind, some scholars believe a genuinely continuous tradition of marking is represented, while others see them simply as sporadic recurrences of simple motifs.

In any case, only the Upper Palaeolithic has so far produced definite figurative , designs, and these become particularly abundant and impressive in the Magdalenian. The technique of 'champlevé' was invented, where bone around the figure is scraped away, making the design stand out as in a cameo (see Fig.40); and the skill was developed of engraving on bone shafts and on batons of antler, not only lengthwise (there are numerous compositions involving lines of animals, or heads) but also around the cylinder. Here, amazingly, perfect proportions were maintained, even though the whole figure could not be seen at once: the finest examples include that of Lortet, with its deer and salmon; that of Montgaudier, showing seals and other figures; and the gannet bone from Torre, Spain, with its fine collection of human and animal heads (Fig.42)[68]

– all decorated bird bones belong to the Magdalenian, as do most depictions of birds.

A certain number of regional differences in these engravings have been apparent since Piette's time[69] (although inevitably there are exceptions to the rule): Dordogne specimens often have very deep incisions, almost bas-reliefs, whereas in the Pyrenees and Cantabria engravings are generally finer, with a mass of detail provided of the animals' coats. Scenes or 'tableaux' are also more common in the Dordogne than elsewhere.

It should also be noted that the art of engraving on teeth, already seen in the simple motifs incised on pendants, was further developed in the Magdalenian, as in the series of bear canines from Duruthy (Landes), with their engravings of a seal, a fish, of 'harpoons', etc.[70]

Carved bone and antler

As mentioned earlier (p.26), a carved and engraved bone is known from the Pleistocene of Mexico; but once again it is Europe, and particularly the Magdalenian of France, which can boast the finest and most numerous specimens of the art. Apart from the perforated and sectioned bones mentioned above, one of the earliest examples is the 'phallus' from abri Blanchard, carved in a horn-core. In the Gravettian, tools and weapons began to be decorated with both figurative and geometric motifs,[71] though this practice really came into its own in the Magdalenian – for example, almost all portable decoration in Cantabrian Spain comes from this period, and increases through the period. However, it should be noted that much of the simplest 'decoration' of tools and weapons – particularly transverse incisions near the base – was probably intended to strengthen the adherence to the haft, no doubt aided by gum or resin, and to improve the grip of the user: it has been called 'technical aesthetics'.[72]

Presumably, decorated objects were for long-term use, since there is little point in investing time and effort in engraving an implement which can be easily lost or broken – but this is not an absolute rule, since some sites, such as abri Morin, have 'harpoons' and spear-points with carefully made, figurative engravings.[73] It is worth noting that Garrigou interpreted the small, perforated Azilian harpoons of La Vache (Ariège) as ear-pendants! Semi-cylindrical rods of antler were carved with a variety of motifs, such as the well-known 'spiral' decoration found in a cluster of Pyrenean sites; perforated antler batons were also decorated, especially in the Magdalenian. Antler, unlike bone, is relatively easy to engrave.

In the mid-Magdalenian, one encounters figures of animals and fish ('contours découpés' or 'perfiles recortados')[74] and circular discs ('rondelles' or 'rodetes')[75] cut out of thin bone. The discs were often cut from shoulder-blades, and several examples of the latter are known with circles removed from them (see Fig.44); many are engraved, either with animal or human figures or with abstract designs like sun-rays, and some have tiny perforations round the circumference (Fig.43). Those with a central perforation have occasionally been interpreted as buttons, which seems an unlikely function, in view of their fragility. Similar discs are also known in other materials: for example, the grave of Brno II (Czechoslovakia), dating to c 23,000 BC, had specimens not only in bone but also in stone, ivory, and cut/polished mammoth molar,[76] while Gönnersdorf has some in slate.

Animal figures are occasionally large, such as the 22 cm bison from Isturitz, found in two pieces 100 m apart, and probably cut from a pelvic

bone;[77] but the majority of the (approximately) 150 'contours découpés'
known are animal heads (about two-thirds of them horses) cut from a horse
hyoid (bone of the tongue), the natural shape of which already bears some
resemblance to a herbivore head.[78] Many are perforated – some through
the nostril or eye, presumably for figurative effect, and others probably to
serve as pendants – and they have differing degrees of detail engraved on
them: eyes, muzzle, coat, and so forth (Fig.45).

Almost all of them have been found in the French Pyrenees, although
sites in Asturias and Cantabria have recently produced a few very fine
specimens of exactly the same type. No doubt the most outstanding find,
hidden in a corner of the cave of Labastide (Hautes Pyrénées), is what
seems to be a necklace of 19 identical perforated heads, apparently of the
isard (Pyrenean chamois) with its cold-season markings, together with one
perforated bison head, all cut from horse hyoids;[79] this remarkable
ensemble was clearly made by a single artist.

It is worth noting that, like portable art as a whole, the distribution of
'contours découpés' is extremely uneven: of those in France, about two-
thirds come from three Pyrenean sites (the 'supersites' Mas d'Azil and
Isturitz, plus Labastide because of its necklace); if those from Arudy are
added, it means that over 75% come from only four sites.[80] Bone discs are
rather more widespread, but over half of them come from the Pyrenees,
with Isturitz and the Mas d'Azil again the richest sites.

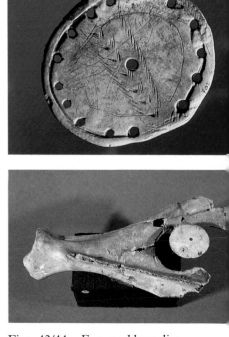

Figs. 43/44   *Engraved bone disc,
with a central perforation and 16
others around the edge, from Le Mas
d'Azil (Ariège). The engraved motif
may be non-figurative.
Magdalenian. Maximum diameter:
5 cm. (JV). Below: a shoulder-blade
from which a bone disc has been cut.
Le Mas d'Azil (Ariège).
Magdalenian. Total length: 24 cm.
(JV)*

The antler spear-throwers of the Magdalenian tend to have two kinds of
decoration: animal heads or forequarters carved in relief along the shaft (a
type found in both France and Switzerland); or figures carved in the round
at the hook-end of the object, where the roughly triangular area of
available antler dictates the posture and size of the carving.[81] However,
within these constraints the artists produced a wide variety of images –
fighting fawns, a pheasant, mammoths, a leaping horse and so forth
(Figs.18,103).[82]

Many of the finest of these carvings have been found in the Pyrenees,
and none finer than the intact spear-thrower from Mas d'Azil with its
image of a young ibex which stands, turning its head to the right and
looking back to where two birds are perched on what seems to be an
enormous turd emerging from its rear end (see Fig.46); this composition is
all the more startling because of the almost identical specimen found a few
years later at Bédeilhac, a few miles away – this one had lost its shaft, and
the ibex is kneeling and turns its head to the left, but otherwise is identical
in all respects. Broken specimens have also been tentatively identified
from other sites in and near the Pyrenees, with the result that up to ten
examples are known; if one allows for preservation, recovery, recognition
and publication it becomes obvious that these must represent a tiny
fraction of the dozens – perhaps hundreds – originally produced. One can
therefore argue for a high output by an individual artist or a small group of
artisans with a favourite theme, since all the examples are attributed to the
Middle Magdalenian, a period which spans a few centuries in the
Pyrenees.[83] Even more remarkable is the virtually identical pose struck by
Bambi and two birds (though without the turd!) in the Disney cartoon
made before the two intact spear-throwers were discovered.[84]

Fig. 45   *Bone 'contour découpé' of a
horsehead, from Enlène (Ariège).
Magdalenian. Length: 5 cm. (JV,
collection Bégouën)*

Statuettes and ivory carvings

The simplest free-standing figurines known from the Upper Palaeolithic
are terracotta models; their existence originally came as a surprise, since

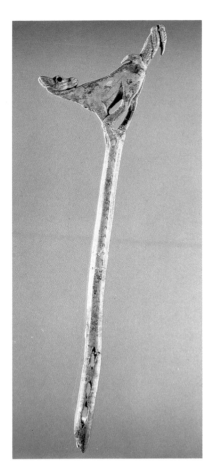

Figs. 46/47 *Antler spear-thrower showing a young ibex with emerging turd, on which two birds are perched. Le Mas d'Azil (Ariège). Magdalenian. Total length: 29.6 cm. (JV). The detail shows the other side of the ibex. Length: c 7.5 cm. (JV). The figure resembles a pose in Walt Disney's 'Bambi', featuring the fawn with two birds: this cartoon feature was released to movie theatres in 1942. The Mas d'Azil spear-thrower was discovered between 1939 and 1941, but the first illustration of it was published only in 1942 (by M. & S-J. Péquart in 'la Revue Scientifique' vol. 80, Feb. 1942); the authors had not yet seen 'Bambi' but made a comparison (pp. 94/5) with the fawns in Disney's 'Snow White' of 1937. In a subsequent publication (1960/3, p. 299) they also refer to 'Bambi' and reveal that illustrations of the carving and of Bambi had been placed side by side in the archives of the Louvre as a comparison!*

one of the dogmas of archaeology was that fired clay was invented later, in the Neolithic period, and that the people of the Ice Age were incapable of making it. This was clearly nonsense, as shown by the mastery of clay in that period (see below, p.93) and by the fact that any fire lit on a cave-floor will have hardened the clay around it: indeed, lumps of fired clay around hearths bear a marked resemblance to crude potsherds. The recent discovery that pottery in Japan dates back at least 12,000 years merely underlines the fact that, if Ice Age people did not make pottery, it was through lack of need rather than through ignorance.

Nevertheless, a number of terracotta figurines have survived and been recovered in different areas: a few examples are known from the Pyrenees, North Africa, and Siberia;[85] but considerable quantities – 77 fairly intact, together with over 3,000 fragments (and others poorly fired, which have disintegrated) – have been found in Czechoslovakia, at the open-air sites of Dolní Věstonice, Pavlov and Predmost, where they are securely dated to the Gravettian, c 22,400 BC. They comprise small figurines of animals and humans, and display some spatial differentiation (ie herbivores in one hut, but human figurines in the centre of another, together with carnivores); a hearth or 'oven' for their manufacture has also been found.[86] The best known of these figurines is the 'Venus' of Dolní Věstonice, made of a mixture of clay and bone powder.

The Upper Palaeolithic also produced carvings in other materials such as soft stone, as at Isturitz and Bédeilhac (figurines which seem to have been deliberately and systematically broken),[87] and occasionally in steatite, coal, jet or even amber;[88] there are also great numbers of '*pierres*

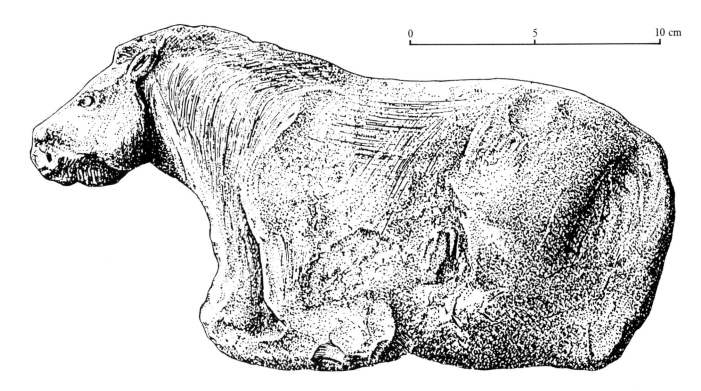

figures', entirely natural objects which bear a fortuitous resemblance to something figurative – it is an open question as to whether this resemblance was noticed or considered significant by Palaeolithic people.[89]

Sculpture is almost impossible in bone, which is hard and fibrous. Consequently, the great majority of figurines are in limestone, sandstone or ivory. The rock-shelter of Duruthy, Landes, has yielded fine horse carvings in each of these materials (see Figs.48,51),[90] while the cave of Lourdes produced the best-known ivory horse. Human figures were carved out of horse teeth at Mas d'Azil and Bédeilhac.

The famous 'Venus' of Willendorf is made of a particular type of oolitic limestone not found in Austria and was thus brought in from elsewhere.[91] Let us note that this figurine, like a few others, still bears traces of red ochre. Stone figures were presumably carved with powerful flint tools, and some traces of the process can occasionally be seen, as on the 'Venus' of Tursac, made from a pebble.

Mammoth ivory was used quite extensively in the Ice Age, not only for statuettes but also for beads, as we have seen, and for bracelets and armlets such as those found on the Sungir skeletons, or the various objects of Mezin (USSR) with their rich decoration of chevrons, zigzags and other geometric motifs.

Ivory is easy to engrave along its fibres, but not across them; a kind of '*champlevé*' could be achieved, as in the Aurignacian human figure with raised arms from Geissenklösterle, Germany.[92] However, three-dimensional figurines were mastered in this material at a very early stage: indeed, some of the earliest known pieces of Ice Age art are ivory statuettes, including the human head and female figures from Brassempouy (Landes), the animal and human figurines from Vogelherd (Germany), and those of Geissenklösterle (fragments of two mammoths and a feline).

Fig. 48   *Drawing of a stone statuette of a kneeling horse, Duruthy (Landes). Magdalenian. (After Laurent)*

Scholars have usually assigned the Brassempouy statuettes to the Gravettian, simply because one or two 'Venuses', such as that of Tursac and perhaps that of Lespugue, came from a layer of that period;[93] among the very few 'Venuses' in western Europe with a stratigraphic context, there are also specimens from the Magdalenian (eg Angles sur l'Anglin[94]), and consequently all 'Venus' figurines have been unjustifiably lumped into either the Gravettian (for the most part) or the Magdalenian. It is extremely difficult to discover from Piette's excavation reports precisely where the Brassempouy statuettes came from (assuming that those he found came from the same layer as those dug out like potatoes by earlier workers); but Breuil, who visited Piette's dig in 1897, saw the relevant layer, and kept samples of the associated industry, stated quite clearly on several occasions that the figurines came from the earliest Aurignacian, and perhaps even the Châtelperronian.[95] In short, Palaeolithic figurative art produced some of its greatest masterpieces – the Brassempouy head, the Vogelherd animals,[96] and the male statuette from Hohlenstein-Stadel, Germany (Fig.49),[97] which dates to *c* 30,000 BC – in its initial phase, which suggests strongly that they must have been preceded by a long tradition in carving materials which have not survived.

Experiments have been carried out in carving 'Venuses' in ivory as well as in stone and other materials; to make an ivory figure, a piece of tusk was first prised out by cutting two deep grooves into the material; this was then

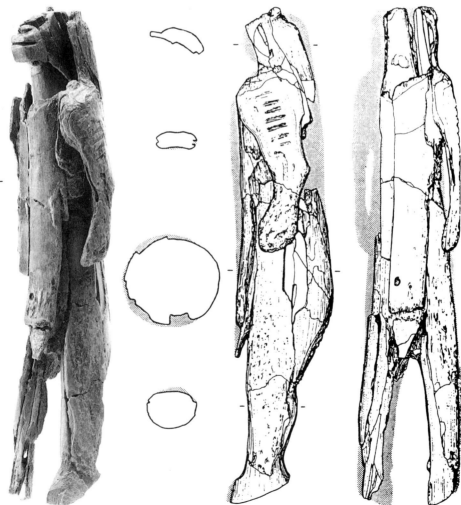

Fig. 49 *Ivory statuette of a human with a feline head, from Hohlenstein-Stadel (Germany), together with drawings of the figure before the face was fitted (after Hahn). Aurignacian. Height: 28.1 cm. (Photo Ulmer Museum)*

rubbed into a rough shape with sandstone, and a burin and broken blade were used to carve the legs and other areas. It is thought that long handling of the figures will have smoothed away any striations left by the work.[98] Carving ivory is very hard, and takes many hours, even days, of effort.

Ivory statuettes were also made in composite form, as in the case of the Czech male figurine from Brno II, dating to c 23,000 BC, whose head, trunk and arm were fitted together to make an articulated 'doll'.[99]

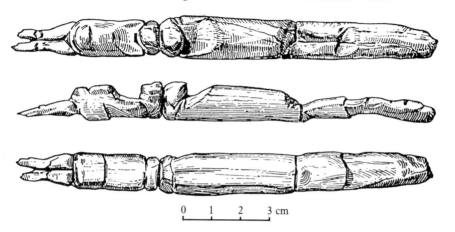

0   1   2   3 cm

Fig. 52   *Figure of a human drawn around a natural phallus-like stalagmite protruding from the wall in Le Portel (Ariège). Probably Magdalenian. Height: 38 cm. (JV)*

Fig. 50   *Two ivory figurines still joined at the head, from Gagarino (USSR). Probably Gravettian. (After Tarassov)*

Some idea of how the human figurines of the Soviet Union were made is provided by the finds at Kostenki and other sites which are so numerous that there are examples of every stage of manufacture, from rough-outs to final polishing;[100] one piece of ivory recovered from Gagarino, a site which has yielded eight ivory 'Venuses', has been carved to form two human figures of different lengths, attached at the head like Siamese twins.[101] Although it is possible that this may have some special significance (especially since the two children at Sungir were buried head to head), it seems more likely that the piece is simply two separate figurines, made together, which have not yet been separated (Fig.50).

It is worth noting that there are regional differences in the material used for the 'Venuses' and in their locations: in the Pyrenees, eastern Europe and Siberia they are predominantly of ivory, while those of Dordogne are in stone, those of Grimaldi are steatite, and those of central Europe employ ivory, stone and terracotta.[102] Most western specimens have been found in caves or rock-shelters, while almost all those from central and eastern Europe come from open-air settlements, and seem to have had a special role in the home (see below, p.140).

As mentioned above, many of those from western Europe, including most Italian specimens, have no context, but the dated examples are scattered through the entire Upper Palaeolithic. It is therefore a somewhat pointless exercise to treat them as a homogeneous or contemporaneous group, as has often been done;[103] but, treated simply as a category of object, like spear-throwers or jewellery, they have a strange distribution: a virtually uninterrupted spread over 3,000 km from Brassempouy (Landes) to the numerous specimens from Kostenki and other Russian sites; and then a 5,000-km gap between Kostenki and the Siberian sites (such as Mal'ta) around Lake Baïkal.

There are many such enigmas in the regional variations and the distribution of different types of portable art. In some cases, it is the availability of materials which is the cause: central and eastern Europe clearly had great quantities of mammoth bones, which were often used for

Fig. 51   *Horsehead carved in limestone, from Duruthy (Landes). Magdalenian. Length: 7.1 cm, height: 6.2 cm. (PGB)*

the construction of dwellings,[104] and this helps explain the dominant use of ivory for carving; the raw material's properties, shape and volume will, in turn, have determined the nature and extent of the decoration or images. Some materials (ivory, antler, certain stones) were suitable for sculpture, others (bone, stone, antler) for engraving. In some cases, the shape placed clear limits on the composition – figures on spear-throwers have already been mentioned, and the same is true of antler batons, long bones or ribs which tend to have linear arrangements of figures. Similarly, hyoid bones were merely made to look more like animal heads than they already do, which explains why some '*contours découpés*' are slightly too long or short to be truly naturalistic.

The basic question, of course, is whether the composition was chosen to fit the support, or vice versa: this is particularly pertinent to plaquettes

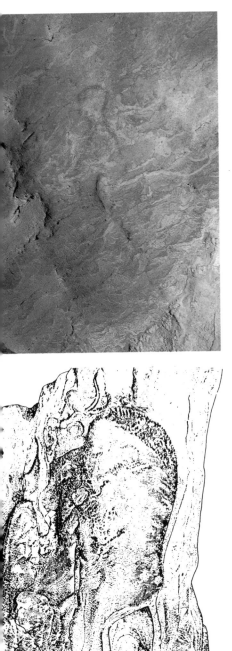

Fig. 53 *Vertical bison painted at Castillo (Santander), utilising the shape of a stalagmite for its back, tail and hind leg (with explanatory drawing, after Ripoll). Probably Magdalenian. Length: 80 cm. (JV)*

Fig. 54 *Deeply engraved motifs on limestone block (60 cm high) from abri Cellier (Dordogne). The principal oval is c 10 cm long. Aurignacian. These motifs have often been interpreted as vulvas. (JV)*

which, unlike bones, come in a wide variety of sizes. Was a composition planned, and then a stone of suitable size selected? Or was the stone chosen first, and the figures made to fit? No doubt, both scenarios occurred frequently, and it is sometimes impossible for us to tell. In a few cases, however, reconstruction has cast light on the problem – at Gönnersdorf, for example, some big slates, more than a metre in length, have been reconstituted from over 50 fragments; on one of them, a large horse figure could be seen only after the reconstruction, whereas some figures were engraved on fragments, their form and size respecting the borders of the plaquettes. Clearly, these slabs were being engraved at every stage of the breakage process, and the figures corresponded to the size and shape of the slates available.[105]

Almost all the Gönnersdorf figures are perfect, with only a few badly proportioned 'failures'. Occasionally, however, we can see apparent mistakes in the engraving process (as distinct from multiple versions of some features; see below, p.184): for example, a pebble from Mas d'Azil bears an engraving of a horse whose body turned out to be too big for the stone, so that the head's position had to be altered accordingly.[106] In other cases, when a drawing proved too big, it was continued on the sides and even on the other side of the stone. If nothing else, slip-ups of this kind show that our ancestors were all too human, and that Palaeolithic art does not consist exclusively of masterpieces!

Intermediate forms – 'pseudoparietal art'

As mentioned earlier, the division between portable and parietal art is simply one of convenience, since there is an area of overlap principally comprising blocks of stone which, although movable, could not be carried around and which may in some cases be fallen fragments of decorated wall. All are from well-lit rock-shelters.

Most of the best-known blocks are thought to date to the Aurignacian although, if they fell from the walls on to Aurignacian layers, they could be considerably older.[107] Certainly the decoration of blocks is known at the

Mousterian site of La Ferrassie (Dordogne), where a Neanderthal child burial lay beneath a large limestone rock with a series of small cupmarks (mostly in pairs) carved in it and apparently placed at random,[108] yet another example of the wide variety of symbolic behaviour in this early period.

Some Aurignacian blocks have deep engravings (almost bas-relief), others have paint on them. The best-known painted specimen is that from the abri Blanchard, where the rock was painted red, and then an animal figure (the bottom part of which survives) was painted on top, with a black outline and a reddish-brown infill.

Occasionally – as in the case of an Aurignacian block from Ferrassie, where the 'broken side' has been painted, not the smooth side which formed part of the shelter vault[109] – it is clear that the artwork was done after the fragment fell from the wall; but by and large one cannot be certain. However, there are so many painted chunks of rock in these early collapsed shelters that it seems highly probable that some of the back-walls and vaults must have been painted (and perhaps engraved) over wide areas – the fact that nothing remains on the walls themselves is simply an indication of how remote this art is: around 30,000 years old, more than twice as ancient as most surviving Palaeolithic parietal art is estimated to be.

Observation and experimentation show that some of the blocks were prepared by grinding and rubbing to produce flatter and better surfaces for engraving.[110] Fine engraving is rare – most of the incisions are deep and wide, some of them clearly made by crudely joining together rows of cupmarks or hammer-blows, and all very weathered.

Brigitte and Gilles Delluc have devised a system of representing the different techniques used on these blocks – as well as in parietal deep engravings and bas-reliefs – in cartographic form, with different symbols and conventions to indicate the depth of the incisions, their shape in section (curved, angular, etc) and so forth.[111] Such 'maps' avoid many of the problems mentioned in an earlier chapter concerning copies of the art, and at the same time convey a great deal of information about the figure's three dimensions.

Parietal art

The principal advantage of parietal art over portable is that, whereas the latter may have been made far away from its final resting place, and indeed in a period earlier than the layer in which it is found, wall-art is still just where the artist placed it. As will be seen in a later chapter, this simple fact provides us with a great deal of information. Like portable objects, parietal art encompasses an astonishing variety and mastery of techniques.

Slightly modified natural formations

In a few cases, it is hard to tell whether the use of natural features was intentional or not, but this is usually pretty clear:[112] for example, in the bosses of the Altamira ceiling used as bulging bison bodies (Fig.7); the shape of cave-walls and stalagmitic formations incorporated into animal backs and legs at Cougnac (Fig.89), a bison back and tail at Ekain, or a bird at Altxerri; the little stalagmites turned into ithyphallic men at Le Portel (Fig.52); the antlers added to a hollow at Niaux which resembles a deer head seen from the front (Fig.55), or, in the same cave, the engraving of a

Fig. 55 *This fissure in the Salon Noir at Niaux (Ariège) looks something like a deer head seen from the front, and the artist added two antlers. Probably Magdalenian. Total length: c 40 cm. (JV)*

Fig. 56 *Detail of one of the engraved owls at Trois Frères (Ariège) [see Fig. 92], showing the striations in the lines. The head is about 10 cm wide. (JV, collection Bégouën)*

bison on the clay floor which began with, and was composed around, an eye formed by a cupmark left by water droplets There are countless other examples.

Finger markings

The simplest form of marking cave walls was to run one or more fingers over them, leaving traces in the soft layer of clay or '*Mondmilch*' (a white, clayey precipitate of calcium carbonate): as shown by the recent finds in Australia (p.28), the technique is extremely ancient, and may have been the first used: it is certainly a method which requires neither great effort nor any kind of tool. Finger-markings in the European caves may well span the entire Upper Palaeolithic and even part of the Mousterian; most scholars followed Breuil in attributing them to a very early phase, but those of caves such as the Tuc d'Audoubert and the Réseau Clastres are almost certainly Magdalenian.

It is possible that in some cases the *Mondmilch* was actually removed in this fashion – either for body decoration or for medical purposes, since in some Alpine regions in historic times it was used as an ophthalmic analgesic[113] – but more often the fingermarks are splayed, and the *Mondmilch* compressed rather than removed, suggesting that marking was the aim.

In Australia, the lines made by fingers seem to be purely non-figurative; but in some European caves, such as Gargas and Pech Merle, they also include definite animal and anthropomorphic figures (Fig.59).[114]

Engraving

One possible source of inspiration for the finger technique is the abundance of clawmarks of cave-bear and other animals on the cave-walls; in a few cases these seem to have been incorporated into designs, such as a circle at Aldène (Hérault), or a hand at Bara-Bahau (Dordogne) (a very similar, well-carved hand is known in Koongine Cave, South Australia);[115] and it is thought that some marks on cave-walls, at La Croze à Gontran (Dordogne) and elsewhere, are engraved imitations of these clawmarks.[116]

Engraving itself, which, as in portable art, is by far the most abundant technique on cave-walls, encompasses a wide variety of forms, the choice of which was largely dictated by the nature of the rock; incisions range from the fine and barely visible to the broad, deep lines already seen on the Aurignacian blocks; scratching and scraping were also used at times.

The deep engravings on the walls of Pair-non-Pair seem to have been done with stone picks, whereas the finer incisions of later periods can be attributed to sharp instruments producing a fine, V-shaped section. Jean Vertut was developing new techniques to visualise the profile of fine engraved lines, using microphotographic analysis and even computer enhancement of contrast and equidensities.[117] It will be recalled (p.42) that his macrophotography reveals the very slight damage done by direct tracing methods, which affects our ability to assess the tools used in the incisions.

A robust tool must have been used for the deep engravings and bas-reliefs in the caves of Domme (Dordogne), and therefore the flint burin found in a narrow fissure in the Grotte du Pigeonnier here, in the immediate proximity of the parietal figures, may simply have been used for initial sketches, or for final details.[118] In any case, such finds are not necessarily always of relevance to the art: for example, a burin-scraper

Fig. 57   Engraving of a vertical horsehead in clay, Montespan (Haute Garonne). Probably Magdalenian. Length: 15 cm. (Photo C. Rivenq)

Fig. 58   'Kneeling' reindeer engraved and scraped on a wall in Trois Frères (Ariège). Probably Magdalenian. Length: c 20 cm. (JV, collection Bégouën)

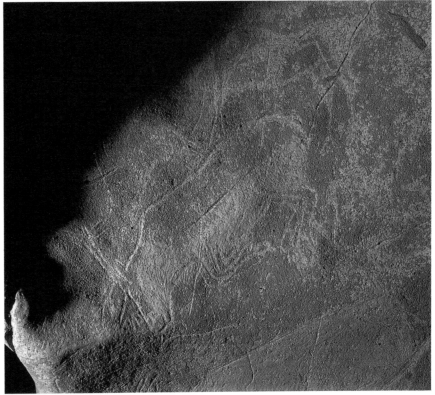

found in a fissure between the legs of the engraved feline of Trois Frères was traditionally assumed to be the artist's tool, left there when the job was done, but in fact nine other objects (flints, bones, a tooth and a shell) have also been found in fissures in this little 'sanctuary'.[119]

As in portable art, the burin was not the only possible tool for the engraver: almost any sharp flint could have been used, from carefully made retouched pieces to simple waste flakes. This is seen clearly at Lascaux, where use-wear was found on the sharp angles of 27 stone tools, including burins, backed bladelets, and simple blades and flakes.[120] Experiments showed that a similar wear and polish could be produced by passing a tool up to 50 times across the rock, thus making an incision 1 mm or 2 mm in depth. These tools were found only in those zones of Lascaux which have parietal engravings; they were absent in parts which do not. A few engraving tools were also found at Gouy, though their use-wear was less pronounced than on the Lascaux tools, no doubt because the rock is softer.

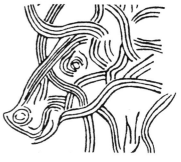

Fig. 59 *Digital tracings at Altamira (Santander) with depiction of bovine head, and tracing by Breuil. The head is just over 1 m in length. (JV)*

It will be recalled that fresh incisions on bone are almost invisible unless the surface is covered with ochre; a similar suggestion has been made for parietal engravings: namely, that the wall may have been coated with a substance such as clay or blood so that new lines stood out from those beneath.[121] It is the great numbers of engraved figures one on top of another at Trois Frères or Lascaux which have led to hypotheses of this type, but in fact they are exceptions to the rule: by and large, parietal figures, whether engraved or painted, carefully avoid superimposition: even at Lascaux, juxtaposition is far more common than overlaps, while at Niaux there seems to be a positive concern to avoid pre-existing figures.

In most cases, fine engravings are almost invisible when lit from the front but 'leap out' when lit from the side. This fact is of some importance, for it provides an indication of whether the artist was right- or left-handed. Right-handed artists tend to have their light-source on the left, to prevent the shadow of their hand falling on the burin (or brush), and accordingly the majority of Palaeolithic parietal engravings are best lit from the left. Occasionally, however, one comes across the work of a southpaw – for example, the Pyrenean cave of Gargas has many engravings, including a fine, detailed pair of front legs of a horse; these had been known and admired for decades; but recently, visiting scholars lit the figure from the right instead of the left and were suddenly confronted with the rest of the horse, which nobody had seen before![122] It is possible, of course, that Palaeolithic artists used the lighting of engravings to their advantage, making them appear or disappear to great effect – alas, we shall never know.

In places where walls were too rough for fine incisions, engraving became almost a form of painting: it was either done more crudely, so that it is best lit from the front, or it was done by scraping. In both cases, the image came not from the relief and light/shadow of an incision, but from the difference in colour between the white engraving and the darker surrounding area of panel. This was the technique used on the Trois Frères feline, mentioned above, and on many other figures.[123]

There are also different regional/chronological styles of engraving. The best example, one easily recognised and well dated (see p.58), is 'striated engraving', using multiple traces, found in a wide area of northern Spain, both in portable art (Parpalló, Rascaño, El Cierro, Castillo, Altamira, etc) and in parietal (Altxerri, Castillo, Altamira, Llonin, etc).[124] However, it is dangerous to assume that every use of a technique such as this belongs to a single chronological phase.

Work in clay

A different example of regional specialisation is work in clay, which (apart from the terracotta figurines discussed above) is restricted to the Pyrenees on present evidence – presumably because the limestone of most of the Pyrenean caves other than Isturitz was unsuitable for sculpture or bas-relief.

The simplest forms of clay decoration were finger-holes, finger-tracings, and engraving with tools in the cave-floor or in artificially set-up banks of clay. Patterns of dots made by fingertip have been found in the floor at Fontanet (Ariège), and apparently random holes were punched by fingers into the frieze on clay at Montespan (Haute-Garonne), which itself comprises horse figures and vertical lines drawn by finger. One of the finest engravings known is the vertical horse-head which begins this frieze (Fig.57).[125]

Figures and signs were drawn in the floor at Bédeilhac and elsewhere, but by far the best-known cave-floor engravings are those of Niaux,[126] mostly located around the Salon Noir; apart from the bison mentioned above, they include figures such as two fish, horses, an ibex head, an aurochs, and something vaguely resembling a human fist. Drawings in clay at Massat include three heads of isard (the Pyrenean chamois).

Such drawings, for the most part, have survived only under rock overhangs or in recesses where the feet of unwary visitors could not reach them. It is therefore almost certain that originally there must have been many more of them, and in far more sites, although at present they do seem to be limited to the Pyrenees, like other work in clay. Untold numbers of fine figures have probably been obliterated by casual visitors to caves in the days before Palaeolithic art became known; only in caves such as Fontanet, the Tuc d'Audoubert and Erberua, where the utmost care has been taken since their discovery, and visits severely restricted, can one be certain that all existing ephemeral traces of this kind have survived.

Since, as we have seen, Palaeolithic people could model clay into figurines, it is hardly surprising that they were equally capable of parietal modelling. Bas-relief exists in cave-floors, as in the series of little horses at Montespan (some almost in haut-relief, since they rise to 9 and 10 cm in height), and also on banks of clay: the best-known example is in a side-chamber at Bédeilhac, where there were four bison (two of them now destroyed), with other marks and a carefully modelled and accurate vulva nearby, with a small fragment of stalactite inserted at the position of the clitoris (Fig.60).[127]

Montespan has some broken haut-relief figures in clay, including at least one possible feline (thought by some scholars to be a horse), but they are so damaged that they pale into insignificance beside the classic examples of the technique: namely the two bison of the Tuc d'Audoubert (Figs. 61-2), which are about one-sixth normal size (63 and 61 cm long) and placed at the centre of a distant chamber, against some rocks. Around them are marks including heel-prints, a crude engraving of a third bison in the cave-floor, and a small clay statuette of a fourth.

The clay was brought from a neighbouring chamber; some 'sausages' on the ground have traditionally been interpreted as phalluses, or

Fig. 61   *The two clay bison of the Tuc d'Audoubert (Ariège) in their context, as the focal point of a small, low-ceilinged chamber. Probably Magdalenian. The bison are 63 and 61 cm long respectively. (Wide-angled photo by JV, collection Bégouën). [See also Figs. 120-22]*

Fig. 60   *Clay bank in the cave of Bédeilhac (Ariège), showing modelled vulva on the left, and a bas-relief bison about 30 cm in length. Probably Magdalenian. (JV)*

Fig. 62   *Detail of one of the Tuc d'Audoubert bison, showing how the clay was worked with fingers and implements. (JV, collection Bégouën)*

occasionally as horns, but a recent study by a sculptor[128] suggests, far more plausibly, that they are simply the result of testing the clay's plasticity, and the position of palm- and finger-prints on them supports this. Similarly, marks on the haut-relief bison show that they were modelled with fingers and further shaped with some sort of spatula (Fig. 62); a pointed object was used to insert the eyes, nostrils, mouths, mane, etc. It is likely that the cracks in the figures occurred within a few days of their execution, as they dried out. Most of the artist's detritus around the figures seems to have been carefully cleared away.

Finally, one fully three-dimensional large clay figure is known: the famous bear at Montespan. However, although a fully rounded form, in a Sphinx-like posture, it cannot really be classed as a statue – it is simply a bear-shaped headless dummy, 1.1 m long and 60 cm high, over which the hide (perhaps with head still attached) of a real bear was probably draped.[129] This would account for the polish on the clay. Nevertheless, a great deal of work went into shaping this considerable mass of clay (c 700 kg), and its haunches and stomach area are particularly fine (Figs.63-64); its claws were also incised into the clay paws.

Bas- and haut-relief in stone

Just as work in clay is restricted to the Pyrenees, so parietal sculpture is limited to other parts of France such as the Périgord, where the limestone could be shaped in this way. One difference between the two techniques is that clay figures are found (or have only survived) inside the dark depths of caves; sculptures are always in rock-shelters or the front, illuminated parts of caves. The reason for this discrepancy remains obscure, since the artists were clearly capable of working for long periods far inside caves and, as we shall see, had light-sources adequate to the task.

Among the earliest examples of parietal stone-carving are the bas-reliefs of Laussel (Dordogne), including the well-known woman holding a 'horn' in her right hand; this particular specimen was originally carved on a large block of four cubic metres in front of the rock-shelter, which was not, therefore, a 'movable wall';[130] the figure includes a number of techniques – her left side is merely deeply engraved, like the Aurignacian blocks mentioned earlier, while her right is in 'champlevé', since material has been removed next to it which brings her body into relief. The figure seems to have been regularised and perhaps even polished; finally, fine engraving was used for her fingers and for the marks on the horn (Fig.67).[131]

The carvings of Fourneau du Diable (Dordogne), on the other hand, are on a block of only half a cubic metre, weighing less than a tonne, and had therefore been placed precisely where the artist wanted them,[132] while the blocks of Roc de Sers (Charente) had originally been arranged in a semi-circle at the back of the rock-shelter, but subsequently some had rolled down the talus slope.

One of the best examples of haut-relief is the frieze at Cap Blanc (Dordogne) (Fig.69), where figures reach a depth of 30 cm in places. By and large, we do not know precisely what implements were used for work of this kind, although it is a safe assumption that they were quite robust percussion tools (hammers, chisels, saws and abrasives), judging from the traces of impact still visible on and around many sculptured figures. Few of the sites have yielded plausible tools in their fill, but at Laussel some Aurignacian 'picks' have battered ends; the picks of Angles-sur-l'Anglin were mentioned earlier (see above, p.58), and worn flint tools including burins and scrapers were recovered from the Magdalenian of Cap Blanc.[133]

The engraving on the Laussel female has echoes in incised lines on the sculptured blocks at Roc de Sers; far more common, however, are traces of pigment on these carved figures, similar to those on portable carvings of various types. The Laussel woman has red ochre on her whole body, especially visible on the breasts and abdomen, as well as in the groove around her, and some of the other Laussel figures have traces too. In the magnificent sculptured frieze of Angles-sur-l'Anglin (Vienne) many figures (including the male portrait) have traces of paint, and vestiges of red ochre have also been found on the fish in the abri du Poisson, and on the friezes of La Chaire à Calvin (Charente) and Cap Blanc – indeed, at the latter the ochre seems to cover large areas, both on the figures and around them, but can now be seen only under very intense light or when the wall is humid.[134]

Pigment was also used in combination with engraving, as at Pair-non-Pair, where there are many traces of ochre inside the incisions as well as on the areas in relief; but in many sites it is difficult to decide whether the engravings were 'rédone' with colour or whether they were originally done on painted surfaces, thus trapping some pigment in the grooves.

Pigments

Identification and preparation

As soon as Palaeolithic parietal art began to be accepted, at the end of the last century, analyses were undertaken to identify the pigments used: Rivière, the pioneer of La Mouthe, was the first to take samples, in 1898, and had them analysed by Adolphe Carnot, who found that they were iron oxide, with no trace of lead or mercury, and had been applied in the form of a watercolour which had penetrated the rock; later, samples from Gargas and Marsoulas, submitted by Regnault to C. Fabre, were also identified as an iron-oxide paint.[135] H. Moissan did analyses for Font de Gaume in 1902, with similar results.

Since that time, the red pigment on cave-walls has consistently proved to be iron oxide (haematite, or red ochre), not only in western Europe but also in remote caves like Kapova and Cuciulat.[136] Black, on the other hand, is usually manganese dioxide, although recent analysis of pigment in the Salon Noir, Niaux, has revealed that charcoal was used for figures here: the plant cells are visible under the microscope, while the scanning electron microscope has shown that it was a resinous wood similar to the juniper; charcoal may also have been used instead of manganese at Las Monedas.[137] Charcoal marks are known elsewhere, such as at Altamira, and the Grotte Bayol (Gard), but these may simply be torch-marks rather than deliberate drawing.

White was used far more rarely; kaolinite was brought into the abri Pataud (a non-painted site, as far as we know) and may have been used as a white pigment or as an 'extending pigment' mixed with other colours;[138] at Altamira, some white paste found in a shell proved to be a mix of mica and illite,[139] and one nodule from Lascaux is also thought to be kaolin.

It has often been assumed that the Palaeolithic artists also used blues and greens extracted from plants, and that these have not survived on the walls – however, it requires complex chemical treatments developed in modern times to extract these colours, and the plants concerned grow in warm or tropical areas; so, apart from the charcoal mentioned above and the possibility of a little woad, plants were probably not used for pigment.[140] Similarly, copper and lapis-lazuli were unknown as sources. Consequently, Palaeolithic artists had only five basic colours to work with: red, yellow, brown, black and white, but the latter is so rare that the choice was really of four.

The main mineral colouring materials were usually readily available, either casually collected as nodules or exploited from known sources (or even occasionally mined, as we have seen in Africa and Hungary); lumps of them have been found in abundance in some sites, and tens of kilos of them, bearing traces of how they were used, lie unstudied in museums.

At Tito Bustillo, in front of the great Magdalenian painted frieze, there were colouring materials, some of them still in the barnacle shells in which they were mixed. In the far earlier cave-site of Arcy-sur-Cure (Yonne), dating to the very beginning of the Upper Palaeolithic, there was a limestone plate which had clearly been used for carrying ochre.

However, the greatest assemblage of evidence of this kind, and the best studied so far, is that from Lascaux, where 158 mineral fragments were

Figs. 63/64   *The front and right flank of the clay bear of Montespan (Haute-Garonne), about 1.1 m in length. Probably Magdalenian. (PGB). The view of the 'dark side' (near the cave-wall) of the bear shows the left forepaw and shoulder, and the finely modelled belly and haunch. Note the traces of red pigment nearby. (PGB)*

Fig. 67   *'Venus with horn' from Laussel (Dordogne). Probably Gravettian. Height: 44 cm. Although now detached, it should be classed as parietal art since originally it was carved on a block of 4 cubic metres. (JV)*

Figs. 65/66   *The sculptured bas-relief horsehead of Comarque (Dordogne), 70 cm long. Probably Magdalenian. (JV). Detail of the horse's muzzle. (PGB)*

Fig. 68   *Red pigment in the cave of Cougnac (Lot). (JV)*

found in various parts of the cave, together with crude 'mortars' and 'pestles', stained with pigment, and naturally hollowed stones still containing small amounts of powdered pigment.[141] There are scratches and traces of use-wear on 31 of the mineral lumps. Black dominates (105), followed by yellows (26), reds (24) and white (3). It was found that there were sources of ochre and of manganese dioxide within 500 m and 5 km of the cave respectively.

The shades vary considerably: the colour of ochre is modified by heat, and Palaeolithic people clearly knew this, since even in the Châtelperronian of Arcy there were fragments at different stages of oxidation still in the hearths. Yellow ochre, when heated beyond 250°C, passes through different shades of red as it oxidises into haematite.[142]

A further stage in pigment preparation, in Lascaux at least, involved the mixing of different powdered minerals, since unmixed pigments are rare here. Chemical analysis of ten samples produced some surprising results: for instance, one pigment contained calcium phosphate, a substance obtained by heating animal bone to 400°C. It was then mixed with calcite, and heated again to 1000°, thus transforming the mix into tetracalcite phosphate.[143] A white pigment was found to comprise porcelain clay (10%), powdered quartz (20%) and powdered calcite (70%); while a black pigment found in a mortar was charcoal (65%) mixed with iron-rich clay (25%) and a few other minerals, including powdered quartz. Clearly, therefore, the Palaeolithic artists were experimenting and combining their raw materials in various ways. Analysis is now proceeding further, and the scanning electron microscope is being used on tiny chips of some Lascaux paintings in order to study the microstructure of the pigments and compare them with natural samples, in an effort to assess the nature and extent of the prehistoric processing.[144]

A different problem is the binding medium used. In the past it was often assumed that some form of fatty animal product was used for the purpose; however, a series of 205 experiments in two caves has been carried out by Claude Couraud, involving a variety of pigments and binding substances (including fish glue, arabic gum, gelatin, egg white, bovine blood, and urine), and a range of wall-types and degrees of humidity. Observation of the results and deteriorations over three years led him to the conclusion that fatty and organic substances were totally unsuitable binding agents, and fail to adhere well to humid walls. In fact, the only substance which seemed to be good at fixing and preserving the pigments on the rock-face was water – especially cave-water, which is rich in calcium carbonate and which was probably used at Lascaux.[145] It was also found that pigments adhered better if they had been finely ground.

Application

The simplest way to apply pigment to walls was with fingers, and this was certainly done in some caves: for example, the animal figures of La Baume-Latrone (Gard) were clearly painted with fingers, as were the 'serpentiformes' of La Pileta (Málaga), versions in red clay or black pigment of the simple finger-markings or 'macaronis' which also abound in these caves.[146] La Pileta and other caves such as Cougnac also have series of double finger-marks in pigment on or around animal figures.[147]

Experiments suggest that painting by finger tends to produce poor results. Nevertheless, when Lascaux II was being made, it became apparent to the modern artists that some painted lines had been done with the finger in the original figures.

Normally, however, paint was applied with some sort of tool; again, since none has survived, it is experimentation which suggests what was used, although one must always remember that the original artists had years of experience and familiarity with their materials, and acquired 'knacks' which cannot be revealed by modern short-term exercises. Some lumps of pigment are in the form of 'crayons' (there are 19 with use-wear at Lascaux alone), and these may have been used to sketch outlines; however, they do not mark the rock well, and really work only on humid walls. Moreover, they wear down very fast – in tests, 1 cm of length produced a mark of only 15 cm![148] Consequently, lumps of pigment must have been used mostly as sources of powder, which explains the traces of scraping on many of them – and flints with ochre on their edge have been found at Lascaux, Tito Bustillo and other sites, close to the minerals.

As mentioned earlier, mortars and other 'vessels' have been found at Lascaux containing crushed pigment, and similar objects with traces of colour inside are known from Altamira (including some vertebrae), Villars (concave fragments of calcite), Tito Bustillo (barnacle shells) and elsewhere. What we do not always know is whether the powder was made into a paste or a liquid – no doubt it depended on the circumstances and the desired effect, whether a dot, an outline or a flat-wash. In some cases, the paint was probably in liquid form, since it is diffuse, or it has run – for example on some figures of the Frise Noire at Pech Merle – although trickling may also have been caused by wet walls at times.

Whether in paste or liquid form, how was the paint applied? Experiments with a variety of materials produced their best results – ie solid, precise and regular marks – with animal-hair brushes (especially badger); brushes of crushed or chewed vegetable fibre were next best. On the other hand, brushes of human hair were too supple and fragile, while pads of bison fur transferred colour to the rock efficiently, but quickly became flaccid and unusable.[149]

Nevertheless, there are cases where some sort of pad must have been used because other methods were unsuitable. Some surfaces at Lascaux have a cauliflower-like covering (the crystalline calcite coating of the Hall of the Bulls made it unsuitable for engravings), and once figures had been outlined on them – presumably by crayon or brush – the infill was done with hundreds of circular, diffuse spots that join up and give the impression of an even wash. They appear to have been done with a pad and dampened powder: for example, a red cow in the 'Diverticule Axial' has been filled by about 200 such patches. Yet these figures have sharp edges, which suggest that a hide or some such object was placed along the desired line.[150] This pad-and-paste method would certainly have been easier to use on ceilings and inclined walls than brushes and liquid paint. In some lines of big dots in various parts of Lascaux, one can see a change after two or three similar ones, which suggests that the pad or brush held enough pigment for only a few at a time.

In some cases, such as the Covalanas and Arenaza deer (outline figures made of dots) or the famous bison of Marsoulas, composed almost entirely of red dots, each 2–3 cm in diameter,[151] it is hard to decide whether fingers, thumbs, pads or brushes were responsible. As mentioned earlier (p.48), use of infra-red film has revealed that some apparently unbroken lines, such as in a tectiform at Bernifal, were originally made with dots whose

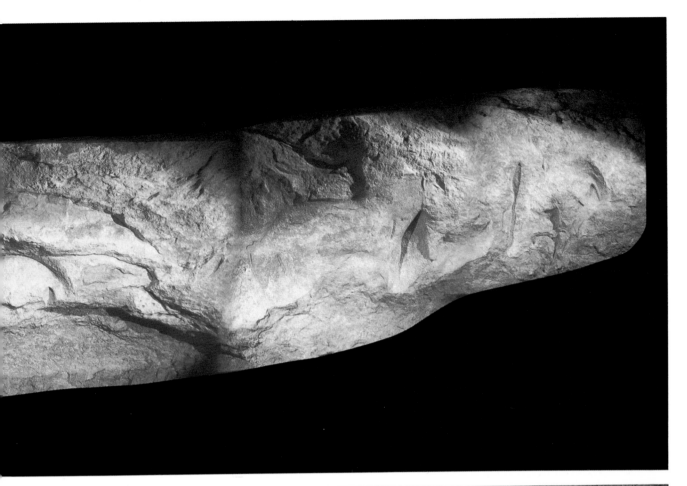

Fig. 69   *Photomontage of the sculptured frieze at Cap Blanc (Dordogne), showing the series of horses, some facing left and others right. Magdalenian. Total length: c 8 m. The central horse is 2.15 m long. (JV)*

Fig. 70   *Three hinds drawn with dots at Covalanas (Santander). Total length: c 2.5 m. The central hind is 60 cm long. (JV)*

Fig. 71   *Some of the incomplete hand stencils of Gargas (Hautes Pyrénées). (JV)*

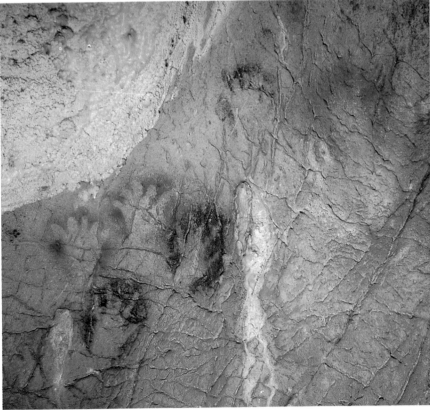

pigment ran together, because the earthy red pigment was soluble in humid conditions and spread (unlike manganese dioxide, which is insoluble and thus does not run and fade). At Combel (Pech Merle), infrared analysis showed that a group of red dots had been made with a pad dipped in ochre, whereas an adjacent set had been applied by spraying.[152]

Hand stencils

A technique of spraying paint was also clearly employed for the hand stencils which are so numerous in certain caves (occasionally with forearm included, as at Maltravieso and Fuente del Salín, Fig. 72);[153] the comparatively rare positive hand-prints (for example at Altamira, Santián, La Pasiega and Fuente del Salín) were made simply by applying a paint-covered palm to the wall. Very occasionally, hand stencils seem to have been made with a pad, as in the case of two black specimens at Gargas which have a regular area of paint around them; a white specimen in the same cave apparently had white material crushed around the fingers; but most have a 'diffused halo' which results from spraying. There are two possible methods: through a tube, or directly from the mouth; and was the pigment in dry or liquid form?

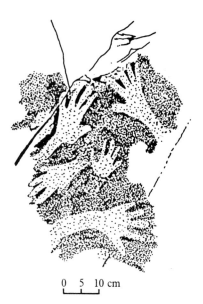

Fig. 72 *Hand and forearm stencils from Fuente del Salín (Santander). (After Bohigas et al)*

Once again, observation and experiments have helped to clarify the situation. When dry, the powder can have been applied only through a tube; the site of Les Cottés (Vienne) contained a bone tube with powdered red pigment inside it, which seems to indicate that this technique was indeed used at times. If dry powder was projected, then a humid wall was required or there would have been no adhesion; in any case, dry pigment leaves 'fallout' beneath the hand, and this does not exist on the Palaeolithic stencils. By contrast, if liquid paint is used, blowing it through a tube concentrates it too much; experiments show that spraying liquid paint from the mouth, about 7–10 cm from the wall, is not only the easiest method but also the one which produces results which best resemble the original stencils;[154] pursing the lips slightly projects a spray of fine droplets which form the required halo with diffuse edges. On average, it takes about 3 g of pigment and between 30 and 45 minutes to do each hand in this way, whereas the 'dry method' can sometimes be quicker, but uses 9–10 g. However, filling one's mouth with paint is hardly pleasant, and it is possible that a combination of the two techniques was used: ie placing a hollow tube upright in a paint-holder, then blowing across its top through a second tube forces the paint up and out in a fine mist; or perhaps blowing into a bent reed so that a spray emerges from a split at its base.[155]

We have already seen that most engravings were probably done by right-handed artists; and it has traditionally been assumed that because most stencils are of left hands (at Gargas, 136 left hands have been identified, and only 22 right), this too denotes predominantly right-handed people. Although this is likely to be correct, the hands are poor evidence, partly because the painting may have been done by mouth, not by the dominant hand, and partly because some hands may have been stencilled palm-upward! However, one should also note that in the very rare Ice Age depictions of people holding objects (particularly those of Laussel, such as the 'Venus with horn') it is usually the right hand which is doing the holding.

Unfortunately, the stencil experiments have not settled the old debate about the apparently mutilated or deformed hands, most abundant at Gargas but also known in other caves such as Tibiran, Fuente del Trucho (interestingly, located just across the Pyrenees from Gargas!), Maltravieso, possibly Erberua and Fuente del Salín, etc. There are three basic explanations for the missing phalanges (finger bones) on these hands: either the fingers were bent (probably palm-upward) when the stencils were done, or the phalanges were actually missing, through ritual mutilation or through pathological conditions.

Many prehistorians have thought the fingers were deliberately bent, for one reason or another, perhaps as a kind of language of gestures or signals,[156] and some experimenters claim to have duplicated all the 'mutilations' at Gargas by bending their fingers, and point out that there are stencils with clearly bent thumbs at Gargas and Pech Merle.[157] Another experimenter, however, consistently found that placing his hand palm-upward and bending the fingers caused pigment to infiltrate behind it,[158] which, as we have seen, was not the case in the originals. He therefore supports the view that the phalanges were actually missing.

If they were, then ritual mutilation may have been responsible, as at Maltravieso, where primarily the little finger was affected; but at Gargas the cause is more likely to be pathological. Investigation by Ali Sahly showed that the stencils here comprised adults, women and/or youths, and children including infants; of the 124 best preserved, only 10 are free of abnormality; the various combinations of missing phalanges are on only the fingers, not the thumb. Almost 30% had the last two phalanges of all four fingers gone. He claimed that repetitions of the same hands were detectable, so that less than 20 people produced 231 prints between them.

His conclusion that conditions such as frostbite, gangrene and Reynaud's disease (which attacks only the fingers and very rarely the thumb) were responsible found support in his discoveries of actual hand-imprints in clay, not only at Gargas but also at Lascaux, which seemed to have phalanges missing; and of finger-holes in clay at Gargas, casts of which seemed to end in stumps rather than finger-tips.[159]

Some painting techniques

Strictly speaking, figures made by finger or in simple outline should simply be seen as coloured drawings; the term 'painting' might better be limited to those cases with infill of various kinds, and, as we have seen, painted infill dates back to the Aurignacian block from abri Blanchard, and perhaps beyond, as does the practice of colouring wide areas of rock-wall. The sophisticated bichrome and polychrome figures of the Magdalenian arrive relatively late, and are rare in Palaeolithic iconography in comparison with engravings and outline drawings.

The latter were undoubtedly quick and easy things to do for an artist of experience and talent. In an already famous experiment, Michel Lorblanchet memorised every mark in the 'Frise Noire' of Pech Merle (there are 25 animals on this panel measuring 7 m by 2.5 m)(Fig. 73), and then reproduced the whole thing on an equally smooth panel of similar dimensions in another cave, lit by a lamp in his left hand. Each figure took an average of 1–4 minutes (compare this with the times for hand stencils, above), so that the whole frieze required about an hour, including initial sketching with a stick.[160]

Similarly, Claude Couraud has reproduced a Lascaux horse and a Niaux bison, both full-size on paper and at reduced scale in a cave: the horse required 30 minutes of preparation and another 30 to do on paper, but took only 30 minutes in the cave; the bison took 20 minutes on paper, and only

10 in the cave.[161] Therefore, it seems that even quite detailed figures needed little time for their execution.

Experiments have also been done to investigate problems of superimposition of different paints. At Lascaux, it was long uncertain whether the red cows were on top of the black bulls, or vice versa, because the two pigments had mixed; some scholars thought the cows were done first, but infra-red pictures suggested the opposite. Using samples of similar pigment, it was found that red on top of black did not mix, but black on top of red did so – and so the bulls were clearly painted after the cows.[162] As mentioned earlier (p.48), infra-red with special cut-off filters

has also been used by Alexander Marshack, Jean Vertut and others to determine the use of different ochres and mixes of ochre within a single composition or panel, where everything looks the same to the naked eye. Similarly, ultraviolet light helps one to separate out overpainting and overengraving more clearly. However, it must be stressed that these techniques cannot be applied universally or at random, but need to be developed for each particular problem or cave.[163]

Combinations of painting and engraving are quite common and varied in parietal art; at Trois Frères, clearly visible traces show that a panel was scraped extensively before a large red 'claviform' was painted on it (Fig. 24);[164] similarly, several panels at Altxerri were scraped extensively to produce a light background for painted figures;[165] in some cases, an engraved figure is filled in with ochre; elsewhere, as in the spotted bison of Marsoulas, a few incisions delimit certain parts of a painted figure, such as the head and leg; while at Tito Bustillo, some of the bichrome horses on the

Fig. 73 *Photomontage of the 'black frieze' in the cave of Pech Merle (Lot), showing mammoths, bison, aurochs, horse and red dots. Solutrean. Total size: c 7 m by 2.5 m. The mammoth at top left is 1.4 m long. (JV)*

main panel are silhouetted by wide areas of engraving which make them stand out from the darker background and from earlier figures.[166] There are even painted animals which are redone or 'highlighted' with engraving, as at Trois Frères, Lascaux, Le Portel, or Santimamiñe, for example.

As mentioned earlier, just as perfectly proportioned figures could be engraved around a cylinder of antler, so the 'falling horse' of Lascaux is painted around a rock: the artist could never see the whole animal, yet its proportions remain sound. It is possible that an even more sophisticated technique was occasionally used in the caves, namely anamorphosis, or deliberate distortion. Claims have been made that four red cows at Lascaux and a black horse at Tito Bustillo were purposely deformed in this way so as to look normal from ground level or from the side;[167] given the other accomplishments of Palaeolithic artists, there is no reason to suppose that touches of this sort were beyond their capabilities.

Just as with portable art, there are clear regional differences in artistic technique (quite apart from style and content): for instance, in the Pyrenees there are roughly equal numbers of engraved and painted figures in the Mas d'Azil, but engraving dominates in the decorated caves to the west and painting in those to the east, although there are inevitably exceptions (Tibiran has more painting than engraving, while Fontanet and Massat are the reverse).[168] The regional difference in distribution between clay and sculpture has already been mentioned; perhaps linked to this is the fact that art on cave-floors is largely restricted to the Pyrenees but, unlike Cantabria and the Périgord, this region has no decorated ceilings – there are occasional engraved or painted figures on ceilings (engravings at Montespan, for example), but there is nothing remotely like the Lascaux passage, or the great painted ceiling of Altamira which Déchelette baptised the 'Sistine Chapel of Rock Art'.[169]

The Altamira ceiling was within easy reach – indeed, it was so low that the artists could not have seen its whole surface at once; but how did they manage to paint on high ceilings? Occasionally, as with a sign in Lascaux, it might have been accomplished with a brush on the end of a long pole. At La Griega (Segovia), it is thought that an engraving on a ceiling 3 m up could have been done by a person straddling the passage with one foot on a ledge on either side.[170] Elsewhere, as at Roucadour (Lot) where the engravings are 6 m above ground-level, or at Baume-Latrone. where there are broad sweeps of painting on a ceiling over 3 m high,[171] it is probable that the floor has sunk down since Palaeolithic times. But for truly monumental work, such as the great Labastide horse on a rock 4 m high, the two black mammoths painted 5 m up on the vault of Bernifal, or much of Lascaux's decoration, it is clear that ladders (perhaps tree-trunks with branch-stumps as rungs) or scaffolding must have been used. Abundant wood residues found in Lascaux may come from these constructions (some are from big oaks) as well as from torches and so forth, but the clearest evidence is the series of about 20 sockets cut into the rock on both sides of the 'Diverticule Axial', about 2 m above the floor; these were packed with clay. Holes about 10 cm deep in this clay suggest that branches long enough to span the passage were fitted into the sockets and cemented into place with the clay. This series of solid joists could then support a platform, providing easy access to the upper walls and ceiling.[172]

Shedding light on the subject

Portable art may all have been done in daylight, like the art on blocks and the decoration of rock-shelters. But the work inside caves required a reliable source of light. In a few cases, a hearth at the foot of the decorated panel may have sufficed: at La Tête-du-Lion (Ardèche) a concentration of pine charcoal only 1.2 m from the decorated wall may be a '*foyer d'éclairage*' of this type, though it could simply be the remains of torches.[173] Certainly, portable light was necessary in most caves. What did they have at their disposal?

Despite Riviere's discovery at La Mouthe, the very existence of Palaeolithic lamps was not fully accepted until 1902, the same year that cave-art was finally validated. This was no coincidence: once the pictures deep inside caves, often hundreds of metres from daylight, were seen to be authentic, it was obvious that the artists must have had some means of illumination.

As often happens, the establishment of a previously rejected notion led scholars to the opposite extreme, so that all hollow objects, and lots of flat ones, were identified indiscriminately as lamps! A recent critical study of the hundreds of objects claimed to be lamps has resulted in only 302 being considered as possibilities, of which a mere 85 are definite and 31 others probable. This is a very poor total when seen against the 25,000 years of Upper Palaeolithic life and the hundreds of decorated caves, especially when it is noted that 70% of them come from open-air sites, rock-shelters or shallow caves.[174] It is certain that a different method was usually employed in deep caves – probably burning torches, which have left little or no trace other than a few fragments of charcoal or black marks on walls.

There are a few beautifully carved lamps; and some, such as those from La Mouthe or Lascaux, even have engravings on them. Many of them are of a red sandstone from the Corrèze region of France, which thus seems to have been a centre of production.

Combustion residues in some specimens have been subjected to analysis which indicated that they were fatty acids of animal origin; while remains of resinous wood or of non-woody material clearly come from the wicks (residues in the carved Lascaux specimen proved to be juniper). Experiments have been carried out with replica lamps of different types, different fuels (cow lard, horse grease, deer marrow, seal fat), and a variety of wicks (lichen, moss, birch bark, juniper wood, pine needles, dried mushrooms, and kindling). The results led to a number of interesting insights, which were confirmed by study of the lamps used by Eskimos.[175]

Firstly, a good fuel needs to be fluid, and easy to light. The initial melting of the fat needs to absorb the wick material which, by burning, continues to melt the fuel. This cycle can continue for hours, providing both wick and fuel are replenished from time to time: one estimate is that 500 g of fat will keep a lamp going for 24 hours.[176] As mentioned earlier (p.23), animal-fat fuel produces no soot.

There are two basic types of lamp: the open-circuit model, in which the fuel is evacuated as it melts; and the closed-circuit, where the fuel is kept in a cavity. The open type seems very rare in the Ice Age, although many simple slabs of stone may have been used like this (about 130 limestone slabs at Lascaux were interpreted in this way, although many are now lost, and only about 36 seem likely lamps); most recognisable Palaeolithic lamps are of closed-circuit type. In Eskimo communities, the open types are for occasional use, while the closed types – in which far more work has been invested – are used daily.

But how bright are these lamps? The answer, surprisingly, is that they are pretty dim, even by comparison with a modern candle. The power of the light given off depends on the quality and quantity of fuel; the flame is usually unstable and trembling. Experiments with a stone lamp using horse fat produced a flame of one-sixth the power of a candle, according to measurements with a photometer.[177] With such limited radiance, it would have been necessary for Ice Age people to use several at once, or to resort to burning torches. In some deep caves, such as Labastide, there are also the remains of hearths of various kinds, which would certainly have provided strong light at times.

We, of course, are spoiled by artificial light, and are no longer accustomed to dimmer sources. But in fact it is surprising how much can be seen by the light of a single candle, and a large cave-chamber could be adequately lit with two or three. It has been found that with only one lamp one can move around a cave, read, and even sew if one is close enough to the light – the eye cannot really tell that the flame is weaker than a candle.

One important factor is that flickering flames of this type have the effect – no doubt observed and perhaps exploited in the Palaeolithic – of making the depicted animals seem to move, an eerie experience. For the really big figures, over 2 m in length, more than one lamp would have been needed to see the whole thing at once. The lamps would also have affected colour perception – they give off a warm yellowish light which makes yellow look orangey, and makes red pale or brown. The eye adapts well to red in this light, and this may explain the frequent use of red dots and signs at various points in the caves – by lamplight they were readily visible signals.

We still have much to learn about how Ice Age people – including many children (p.13) – managed to make their occasional forays into the remote depths of some caves. Even armed with bundles of torches, or lamps with a plentiful supply of fat and wicks, it seems a very risky thing to do. However, if archaeology has taught us anything, it is that we should never underestimate the capabilities of our ancestors. We must not lose sight of the fact that caves and darkness formed part of their environment, and they clearly knew how to cope with them very successfully.

Open-air art

Equally, however, we should not let the terms 'cave-art' and 'cavemen' blind us to the fact that they very rarely lived far inside caves but, rather, occupied rock-shelters, cave-mouths, and a variety of open-air habitations, from the tents of western Europe to the mammoth-bone huts of central and eastern Europe.

Since they spent almost their whole lives in the great outdoors, it has always been assumed that they must also have produced art outside the caves, but that it has not survived the millennia of weathering and erosion. As mentioned earlier, there have been occasional claims that open-air figures were of Palaeolithic age – most notably at Chichkino, Siberia, where hundreds of animal depictions over a distance of about 3 km include a horse and a wild bovid considered characteristic of the end of the Ice Age – but few scholars have been prepared to take them seriously. In recent years, however, a series of important finds in western Europe has finally brought the proof that Palaeolithic people did produce art in the open air, that it can survive, and that the Soviet claims may therefore be valid after all.

The first such finds occurred at a number of rock-shelters and caves (including Murciélagos) in the Nalón Valley, Asturias, which have some eroded lines and animal figures deeply engraved in their exterior areas;[178] nevertheless, these engravings were still in or near shelters and cave-mouths, rather like the sculptured friezes and carved blocks of France. The next discoveries, however, were at truly open-air sites.

At Domingo García, Segovia, the figure of a horse, almost a metre in length, was found engraved (or, rather, hammered out) on a rocky outcrop.[179] In style it resembles the engraved horses in the cave of La Griega, in the same region; moreover, a schematic engraving of a different style and period is superimposed on its outline (Fig.75).

At around the same time, three animal figures, including a fine horse, 62 cm long and 37.5 cm high (Fig.74), were discovered on a rock-face above the River Douro, at Mazouco in north-east Portugal;[180] they had survived thanks to a position which protected them from the elements.

Fig. 74 *Open-air engraved horse at Mazouco, Portugal. Length: 62 cm, height: 37.5 cm (Photo D. Sacchi)*

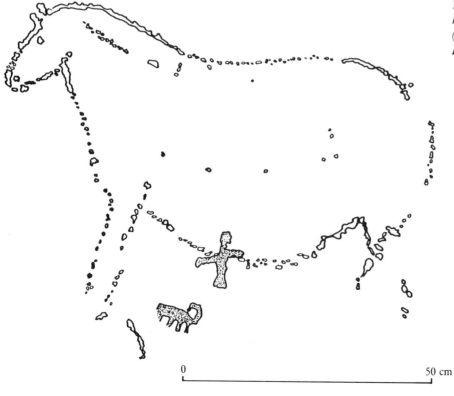

Fig. 75   *The open-air engraved horse of Domingo García (Segovia). (Photo and tracing A. Moure Romanillo)*

0   50 cm

This horse, too, resembles those of La Griega, and the three sites thus seem to form a regional group.

Finally, a series of fine incisions has been found at Fornols-Haut, Campôme, in the eastern French Pyrenees, on a huge block of schist located at an altitude of 750 m on a mountainside (Fig.76).[181] The rock has been greatly weathered by wind; but, because the eastern face is sheltered, its engravings, although eroded, are clearly visible. This face is covered in engravings, drawn in all directions and comprising about ten small animals – none complete – as well as signs and zigzags. The finest figures include the head of an isard, 7.5 cm high, and that of an ibex (Fig.77).

As we have seen in the preceding chapter, all parietal art is notoriously difficult to date, open-air examples especially so since they have no context whatsoever. However, there is nothing remotely similar to these figures in the Middle Stone Age and later art of eastern Spain, or in the schematic art of more recent periods; moreover, had any of these figures been found in a cave they would automatically have been accepted as fine specimens of Palaeolithic art. These sites have therefore proved the existence and survival of Ice Age open-air engraving (painting was doubtless also done outside, but is far less likely to have survived weathering in this way); and they have profound implications for our interpretation of the functions of Palaeolithic art as a whole. Palaeolithic parietal art is by no means limited to caves – they are merely the places where it has been best preserved.

Conclusion: portable versus parietal

In the past, portable art was sometimes seen merely as a 'sketchbook', a repetition or rehearsal in miniature for parietal work. Today, however, we see them as parallel and equally important art-forms, clearly part of the same world, but perhaps entirely different in function.

As we have seen, one basic difference between the two is that portable objects have been moved around, whereas parietal figures are precisely where the artist wanted them, whether on the floor, the walls or the ceiling of the cave. Another difference is that portable art, by definition, is composed of small, light objects, and the figures and motifs drawn on

Fig. 76 *The engraved rock on the mountainside at Fornols-Haut (Pyrénées-Orientales), altitude 750 m. The rock is 2.3 m high and 3.9 m wide at the base. (Photo D. Sacchi)*

them are of a corresponding scale; as we have seen, the size and shape of
the stones, bones and antlers either determined the form of the design
(lines of figures are common) or were chosen according to requirements.

Fig. 77   *Ibex head engraved on the rock at Fornols-Haut. Length: c 7.5 cm. (Photo D. Sacchi)*

    The same applies to parietal art (especially where rock-shape was
incorporated into a figure), but the height and width of the available
surfaces meant freedom from many of these constraints; hence, parietal
figures could be arranged in a variety of ways (individuals, clusters,
panels), and simple lines of animals are rare. Moreover, there was no
restriction on size: figures range from the tiny (such as those of Campôme)
to the enormous (such as the great Labastide horse, almost 2 m long, for
which scaffolding was required[182]). The Hall of the Bulls at Lascaux has
relatively small deer and horses, together with the huge bulls, over 5 m in
length; but of course not all the figures on a panel of this type are
necessarily of the same date, meaning or composition, and this is the
fundamental problem to which we must now turn our attention.

6

What was Depicted?

For convenience, Palaeolithic images are normally grouped into three categories although, as usual, there is a degree of overlap and uncertainty in this division; the categories are animals, humans and non-figurative or abstract (including 'signs'). Each presents its own problems, but the first two share fundamental questions of zoology (identification of the species) and ethology (what are they doing?). Neither question is straightforward.

Animal figures

The vast majority of animal figures are adults drawn in profile; there are a number of possible reasons for this – it is easier than drawing them full-face, particularly if one wants to convey the important features of the animals' anatomy, which are also far clearer on adults than on the young; and it is certain that this was how the animals were observed and identified by Palaeolithic people.

Allowing that we can never *know* what any prehistoric image was meant to depict – only the artist can tell us – it is obvious that most of the animals drawn seem easily 'recognisable' at genus level; but there are great numbers of figures which are badly drawn (in our eyes), incomplete, or (perhaps purposely) ambiguous. Numerous examples can be found of figures which have received markedly different attributions – beaver or feline, fish or porcupine, bear or reindeer, big cat or young woolly rhino, and crocodile, reindeer or horse![1] The same is true in Australian rock-art, where there can be difficulties in deciding between possums and dingoes, or tortoises and echidnas, since figures have few diagnostic features, and Aboriginal identifications can differ from those reached by zoological reasoning.[2] However, in the absence of informants about Ice Age images, zoological reasoning is all we have to go on – comparison with present-day animals, or, in the case of mammoths or woolly rhinos, with their closest living relatives or with frozen carcasses from Siberia – but such comparisons have an inevitable degree of subjective assessment.

It is therefore no surprise that, just as scholars differ on the numbers of figures in a cave, they also tend to produce very different totals and

percentages for each species: at Marsoulas, for example, the three most
recent studies have claimed 21/32/30 bison respectively, 11/32/26 horses,
and 16/18/12 humanoids; at Le Portel, three recent studies claim 33/26/31
horses, 27/23/30 bison, and 6/5/4 humans![3] As with different tracings of a
single figure (see above, p.48), which version should one believe? The
most recent is not necessarily the most reliable.

Another problem is that many authors, including some of the foremost
authorities on Palaeolithic art, admit in the text their doubts about
particular interpretations; but in captions, statistics and tables these
uncertainties often become definite attributions, and thus distort the
results.[4]

A hierarchy of identification should therefore be adopted – definite,
probable and uncertain – and must be maintained consistently throughout
a piece of work. Even so, differences of opinion will remain.[5] Identification
of figures in art can never fully become an exact science, although expert
knowledge of animal anatomy, as exemplified by the remarkable studies of
Léon Pales, can help to clarify many issues.[6] As shown in an earlier
chapter, it is also wisest to work from originals or from photographs,
rather than from someone else's tracings.

Another type of evidence to bear in mind is the range of species present
in the faunal remains of the site or region in question during the Upper
Palaeolithic. Inevitably, however, there are cases where bones are of no
help – for example, nobody denies that the ibex is depicted at Pair-non-
Pair, although no bones of the species have ever been found in Gironde;
similarly, pictures of seals do not coincide with the few remains of the
creature found in Palaeolithic sites; and despite the existence of a few
depictions of mammoth in northern Spain, remains of the animal have
been found at only one or two sites there.

Where complete or detailed pictures of animals are concerned, there is
generally a broad measure of agreement, providing they are well
preserved: no one could doubt the horses of Ekain or the bison of Niaux.
Problems arise where there is a choice of more than one type of animal such
as species of deer, as will be seen below; or where there is an extreme
degree of stylisation in the figure.[7] A figure's size is of no help, since
Palaeolithic art has no scale or groundline, and, as we have seen, tiny
figures are found next to large ones, even of the same species, and
sometimes on the same panel.

Occasionally, complete animals may defy identification; they resemble
no known species, are presumably imaginary, and have been left
undetermined, though some have been given bizarre names: the spotted
'unicorn' of Lascaux with its two straight horns (which has been
interpreted as a Tibetan antelope, a lynx, and even two men in a skin – see
below, p.157); the 'antelopes' of Pech Merle, with their swollen bodies,
long necks and tiny goatlike heads (Fig.78); or the two strange beasts of the
Tuc d'Audoubert, with their shapeless muzzles and short horns.[8]

There are also figures of beasts with extremely long necks at sites such as
Pergouset and Altxerri; and a few images seem to be composites, such as a
deer at Trois Frères which, according to Breuil's tracing, has a bison's
head; a fish at Pindal which has the fins and tail of a tuna on a salmonid
body;[9] and an animal at La Marche which seems to have the body of a horse
and the head of a bovid, while another has the body of a horse or bovid,
together with an antler.[10]

The many incomplete figures offer a different set of problems. Lack of detail in apparently unfinished animals can lead to different views, as in Niaux's Salon Noir, where a sketchy outline of a quadruped, 1.5 m long, above a horse high on one panel has been interpreted as a feline by some scholars, but as a definite horse by others.[11] Fragmentary animals are not usually a problem when the head is present; where the head is missing, as in a bison on the Altamira ceiling or the bears of Ekain (Fig.90) and Montespan which were purposely left headless, the characteristic shape of the rest of the body leaves no room for doubt; but if only part of the body survives, it can be difficult to decide: one such figure at Bédeilhac which lacks both head and forelimbs is generally thought to be a horse. The back legs on the Laugerie Basse engraving called the 'Femme au Renne' are thought by most specialists to belong to a reindeer, though Leroi-Gourhan would prefer it to be a bison in order to fit in with his theories (see below, p.169).[12]

However, it is all too easy to make mistakes with incomplete figures, particularly on fragmentary plaquettes, where much remains uncertain: a piece from Labastide, for example, had the back half of a quadruped whose shape was thought characteristic of a horse, but when later a further piece was found, the figure proved to be a bison![13] Interpretation can also depend on which way up a plaquette is held – for example, an engraving on a specimen from Lourdes was seen only as a horse head by some scholars, but a fine ibex also appeared when it was turned upside down.[14]

Schematisation, which involves reducing a figure to its essential traits and leaving out the rest, is well known in Palaeolithic art, and is perhaps clearest in the isolated neck- and back-lines of horses, bison or mammoths (or, for horses, simply the mane) – for example, at Marsoulas, Cougnac and Pech Merle.[15] Nevertheless, identification of the animal is still possible: in the Réseau Clastres at Niaux, there are three painted bison; the first is complete and 'naturalistic'; the other two are very incomplete, but even without the proximity of the first they could confidently be identified as bison through comparison with complete examples elsewhere: the shape of the back and position of the mane are quite distinctive (Figs.79 and Frontispiece). Such figures appear abbreviated, perhaps a kind of Palaeolithic 'shorthand' in which a part stands for the whole; however, this implies that we know the artist's intention – and it is possible that the figures simply depict a part of the animal, since the shape of the back and head would be fundamental in recognition at a distance.[16]

As will be seen below, there have been many attempts to derive detailed information about different species or 'races' of animals – especially horses – from Palaeolithic figures; all were doomed to failure, precisely because this is art. We cannot assume that the Palaeolithic view of 'realism' was the same as our own,[17] or that they were concerned in any way with presenting accurately the external features of the fauna. There is no justification for creating new species or subspecies on the basis of these images, as has sometimes been proposed: the international rules of zoological nomenclature require that species be defined on the basis of anatomical remains; it is unscientific to attempt it from pictures. For bison or horses, each art-site could present at least one subspecies of its own, but the diversity probably owes more to the artists than to nature.

Similarly, little reliable information can be obtained about the animals' coats – which in any case vary with sex, age, season and climate – because of omissions, distortions and conventions in the figures.[18] The magnificent antlers and horns, and the optimal body growth suggested by the prodigious amounts of fat on some animals, together with the lack of emaciated animals in the art, have been interpreted as proof of the rich environment and long growth season in these parts of Europe.[19]

The age of the animals can almost never be estimated, except in the case of the few juveniles such as the calf with two adult bison on a bone fragment from abri Morin, the calf on a bone disc from Mas d'Azil (with an adult bovid on the other side), or the fawns apparently suckling at does drawn on the wall at La Bigourdane (Lot) and a plaquette from Parpalló.[20] In other cases, one can make only a subjective assessment of the size of the antlers or horns, which may be an artistic convention rather than an accurate record.

The animals' sex is sometimes displayed directly but almost always discreetly – usually on male bison or deer,[21] but rarely on horses for some reason (eg at La Sotarriza). Only a few animals are known which have female genitalia depicted;[22] this may be partly due to the predominance of profile views, in which the vulva is not visible, though this could easily be overcome through twisted perspective if it was important to convey gender. Even udders are rarely shown (eg on one of the curled-up bison of Altamira), even where they could be, as in the 'jumping cow' of Lascaux.

One therefore has to look for secondary sexual characteristics, such as antlers in red deer, differing horn shapes in ibex, or sheer size and proportions in other species – though there is always a potential overlap between females and juvenile males.[23] These secondary features were sometimes exaggerated – particularly deer antlers and bison 'humps' – and thus comparisons of size cannot be undertaken between sites, since the massive mammoth-like humps of some Font de Gaume bison, when placed beside some of the more streamlined, 'boar-like' bison of Niaux, make the latter appear to be the females to Font de Gaume's males.[24]

So much for identification; but what are the animals doing? Many of them are 'motionless', to the point where it has been claimed that they are dead, or at least copied directly from carcasses[25] – primarily because they seem to be standing on tiptoe, and show no sign of weight-bearing in the limbs; these features are not to be found in living animals. The hypothesis also accounts for some of the twisted perspective in hoofs (since the underside of the foot is frequently turned up by a fallen animal), for the fact that nearside feet are sometimes higher than the others, for the prominence of the belly, and for projecting tongues and raised tails. It is, of course, possible to attribute all of these features to muddled memories of living and dead animals,[26] or simply to artistic conventions, but it must be admitted that close observation of carcasses was certainly the artists' principal source of anatomical detail.

One cannot, however, imagine them dragging such dead-weights far into caves to be copied directly! Perhaps some portable depictions are outdoor sketches of carcasses, which were later carried into the caves and copied on to the walls, but this is pure speculation. The fluent, effortlessly drawn and well-proportioned animal figures in parietal art suggest that the artists carried everything in their mind's eye. As Breuil, who used the same technique, said of the Palaeolithic artists: they never took a measurement – they projected on to the rock an inner vision of the animal.[27] Even if the Palaeolithic figures were ultimately based on dead specimens, it is noteworthy that their motionless figures are nevertheless imbued with an impression of life and power.

Fig. 78   *The so-called 'antelopes' of Pech Merle (Lot): this panel is c 1 m long. (JV)*

Some hunting specialists have chosen to interpret many of the features of these images in behavioural terms: for example, the raised tail is seen as a signal of alarm, anger or threat, of submissiveness, or of a copulatory invitation by females[28] – and thus as an illustration of autumn, when such displays are common due to the interactions of rutting behaviour, although if the threat were directed at a hunter it could occur in any season.

There have been two particularly famous examples of zoological deliberation about posture in Palaeolithic images. The engraved reindeer from Kesslerloch (Switzerland) was traditionally thought to be grazing, until it was pointed out that its stance was exactly that of a male in rut – either sniffing a female's track or, more likely, in a position for attacking a rival.[29] The curled-up bison on the bosses of the Altamira ceiling have been described as sleeping, wounded or dying (see below, p.153), or as clear pictures of females giving birth;[30] currently, the dominant view is that they are males, rolling in dust impregnated with their urine, in order to rub their scent on territorial markers – even though one of them has udders, according to Breuil's copy![31] In fact, they may simply be bison figures drawn so as to fit the bosses – they have the same volume, form and dorsal line as those standing around them, but their legs are bent and their heads are down.[32]

'Animated' figures are rare, and although they have occasionally been placed in the centre of a decorated 'panel', they are more usually to be found at the edges.[33] Such movement seems to appear in Solutrean and early Magdalenian times (as at Roc de Sers), but is most common in the middle and later Magdalenian, although this kind of 'realism' never predominates. However, it is Lascaux which has the most abundant and

Fig. 79 *Three bison drawn in the Réseau Clastres (Ariège) in different degrees of completeness [see also Frontispiece]. The most complete, on the left, is 1.14 m long. Probably Magdalenian. (JV)*

varied animated figures; the cave of Gabillou, in the same region and thought to be of roughly the same period, also has lots of movement in its figures.

'Scenes' are very hard to identify in Palaeolithic art, since without an informant it is often impossible to prove 'association' of figures rather than simple juxtaposition. In a few cases, however, this can be done: for example, the Laugerie-Basse engraving, mentioned above, clearly shows a supine woman beneath and beyond a deer. Different versions of one possible scene, depicting what may be a wolf menacing a deer, have been found on portable engravings from the Pyrenees (Lortet, Mas d'Azil), Cantabria (Pendo) and Dordogne (Les Eyzies), which is an impressive distribution.[34] A few other portable engravings such as the 'bear hunt' of Péchialet (Dordogne), or the 'aurochs hunt' of La Vache have also been interpreted as scenes, as has the parietal group of figures in the shaft at Lascaux (see below, p.186), though one cannot be sure that a narrative is involved rather than a symbolic mythology.[35]

In short, therefore, one has to be very careful in making zoological and ethological observations from Palaeolithic images. If one has photographs of animals, one can accurately identify them and assess what they are doing; but here we are dealing with drawings by artists with a message to convey, and using stylistic conventions: nobody would assume, from the spotted specimens at Pech Merle, that the horses of the period had big bodies, small legs and tiny heads; so how reliable are the other features on display? These are not 'Palaeolithic photographs' – we need to allow for convention, technique, lack of skill, faulty memory, distortion, and whatever symbolism and message were involved. Only the most complete and 'naturalistic' figures are really informative, but even the finest Palaeolithic pictures may prove a disappointment to the zoologist. Ice Age images are exciting because they allow us to 'see' the fauna of the time, to put flesh on the bones which we dig up;[36] but we are seeing the animals through the artists' eyes.

Despite the different identifications and percentages of species produced by each specialist, one fact remains clear: the overwhelming overall dominance of the horse and bison among Palaeolithic depictions. In Leroi-Gourhan's sample of 2,260 parietal figures, no fewer than 610 were of horse, and 510 were bison,[37] which thus account for about half between them. No other type of animal comes close, although, as we shall see, other species do dominate at particular sites. The next most numerous include deer, ibex and mammoth, followed by a cluster of rarer species.

The horse
Study of the most important animal in Palaeolithic iconography has, in the past, wasted a great deal of effort on attempts to establish exactly which 'races' were depicted; Piette had a try at this in the nineteeth century,[38] and was later followed by other scholars who distinguished up to 37 varieties![39] However, all such exercises were eventually abandoned; as mentioned above, they were based on the shape and size of the figures, their manes, and the colour and pattern of the coats, all of which were subject to artistic whim – for example, Gravettian figures tend to have elongated heads, while those of the Solutrean and lower Magdalenian tend to have tiny heads! The only firm basis for subdividing horses into different species is palaeontological differences, primarily in the skull and backbone, and it is therefore a matter best left to those working with faunal remains.

More recent studies have avoided this topic, and instead approached important collections of horse figures from a more rigorous knowledge of equine anatomy.[40] At Gönnersdorf, 74 mostly fragmentary horse depictions are known from 61 plaquettes, making this animal dominant among the animals depicted at the site. Horses likewise dominate the animal figures at La Marche, where at least 91 whole or partial specimens have been found on 64 stones; 62.4% of them face right, a percentage similar to that (57–59%) at sites such as Ekain, Lascaux and Les Combarelles.[41] Not one of the La Marche examples can be sexed; and it has been found that, like artists throughout prehistory and history, those of La Marche consistently made the horses' bodies too long[42] – perhaps to give an impression of speed and lightness. Certainly, in Palaeolithic art as a whole, horses strike a number of different poses, which can be interpreted from a behavioural point of view, such as the jumping horse on a spear-thrower from Bruniquel, or the neighing horse head from Mas d'Azil.[43]

One piece of zoological information which is probably reliable in these horse depictions is that of the length of the coat, since present-day Przewalski horses (thought to be among the closest surviving equivalents to the Palaeolithic types) do display striking seasonal differences: hence the shaggy ones recorded at sites like Niaux must be the long pelt of winter, and, interestingly, no distinct 'winter horses' are known in Cantabrian caves.[44] On the other hand, these differences might also denote different climatic phases, as has been suggested for a similar range in bison depictions (see below).

Another frequent feature of Palaeolithic horse figures is the 'M' mark on their side, which denotes a change in colour between the dark hide and the lighter belly area – a shape and change still clearly visible on modern animals such as the little semi-wild 'pottoks' of the Basque country (Fig.114).[45] Stripes also sometimes occur on the horses' shoulders (for example, at Ekain), a feature which Darwin noticed on living specimens, but which the Palaeolithic artists may have exaggerated.[46]

It is certain that the horse was of tremendous cultural importance to people in western Europe during the Upper Palaeolithic – horse teeth and bones have been found carefully placed in Magdalenian hearths in a number of deep Pyrenean caves such as Labastide and Erberua,[47] as well as near hearths at the Magdalenian open-air camp of Pincevent near Paris (despite the fact that reindeer account for almost 100% of this site's faunal remains).[48] At Duruthy (Landes), the carved horses (see above, pp. 84/6) were found in a kind of 'horse sanctuary' dated to the twelfth millennium BC; the biggest, a kneeling sandstone figure, rested against two horse skulls and on fragments of horse-jaw, while three horse-head pendants were in the immediate vicinity.[49]

Finally, it should be noted that the horse may well have been under close control during the Upper Palaeolithic, as suggested by a variety of evidence including some possible depictions of simple harnesses on certain figures, especially one from La Marche (Fig.22).[50]

The Bison

Like the horse, depiction of the bison varied through the Upper Palaeolithic, in terms of shape, size and shagginess; some scholars have argued that not only artistic convention was involved but also perhaps climatic change, so that the heavy, shaggy bison would denote cold phases and the lighter forms would be linked with milder phases[51] – it is a nice

Figs. 80/81 *Perforated antler baton with carving of horsehead with ears raised and a stylised mane; the shaft is also decorated. Le Mas d'Azil (Ariège). Magdalenian. Total length: 20 cm. (JV). Detail of the horsehead. From muzzle to ear-tip it measures less than 4 cm. (JV)*

idea, but a lot of work and more accurate dating would be required to prove that climate was indeed responsible, rather than style or different species/subspecies. As mentioned above, different sites have their own characteristic bison 'construction', seen in an extreme form at Font de Gaume, and attempts to differentiate specific varieties have led no further than similar studies of horses.

The best study of bison figures, albeit of a small sample (13), concerns those of La Marche;[52] an earlier attempt to establish the proportions and rules of construction that characterise Palaeolithic depictions of bison proved of limited value because it was derived from the figures themselves; it is in fact necessary, in studies of this kind, to proceed from precise anatomical markers taken from living animals or at least from photographs of them in profile.[53] Thus nature, rather than artistic convention, is the reference point. Using this method, Léon Pales found that Palaeolithic bison figures, even within a single site (such as Niaux or Fontanet), showed differences in execution and style, whereas other groups (such as at Altamira) were fairly homogeneous. Of all those examined, the La Marche specimens were the 'closest to nature'.

It was also found that in this site, as at Trois Frères and Font de Gaume, there was a tendency to have bovids facing left. Depiction of gender varies

Fig. 82 *Bison engraved and painted in Trois Frères (Ariège). Magdalenian style. Length: 1.5 m. (JV, collection Bégouën)*

Fig. 83 *Apparent scene of bison pursuing a human, carved in bas-relief around a block from Roc de Sers (Charente). Solutrean. The block is c 35 cm thick, the human is 50 cm high, and the bison 54 cm. (JV)*

from site to site: at Font de Gaume, 12 of the 39 bovids have a phallus, whereas none of Niaux's 47 bison has one (although many look male from secondary sexual features such as the hump, the massive head and the short, thick horns); only one at Trois Frères, standing on its hind legs, has a phallus – one of the rare erect phalli in Palaeolithic art – and only one or two of the La Marche bison may have the sex indicated.[54] Anatomical details seem to be incorrect occasionally: for example, horns that point forward, or which differ in volume from those found among faunal remains.

Most bison are 'immobile', but a few do show movement – some hypotheses about the curled-up specimens at Altamira have been mentioned above, and we shall return to them later (p.153). One famous bison, a carving from La Madeleine, has turned to lick its flank; while that in the 'shaft scene' of Lascaux appears to have lost its entrails and perhaps to be charging a man. Other 'scenes' involving bison attacking people may be depicted on the cave-wall at Villars (Dordogne), on a block from Roc de Sers (Fig.83) and a portable engraving from Laugerie-Basse.[55]

The aurochs

Wild cattle are the other large bovids represented in Palaeolithic art; the most lifelike are perhaps those of Teyjat, and the portable engraving from

Trou de Chaleux (Belgium) (Fig.37), but the biggest and best known are those of Lascaux, including the 5-m bulls (Figs.30-2). Nobody has ever questioned the attribution of the latter to *Bos primigenius* but, for some reason, certain leading scholars thought the smaller cattle of Lascaux to be *Bos longifrons* or *brachyceros*, a domestic type with long forehead and short horns which is not dated earlier than the Neolithic period.

The simple explanation is that of the pronounced sexual dimorphism of *Bos primigenius*: the big cattle of Lascaux are bulls, while the smaller are probably cows of the same species. Studies of similar cattle in Morocco with the coloration of their wild ancestors show that the great bulls are black or reddish-black, with a light-coloured strip along the back and some light hair between the horns; the cows are usually reddish-brown, with darker legs, a darker and slimmer head, and thinner horns. All of these features are on display at Lascaux, some to an exaggerated degree, together with other specimens which are slightly abnormal in colour.[56] The sexual dimorphism can also be seen, to a lesser extent, in the bull and two cows engraved at Teyjat (Figs.85-6).

Deer

Apart from a few depictions of the giant deer (*Megaceros*), such as those of Cougnac which cleverly use natural rock shapes for the dorsal line and hump (Fig.89), and sporadic claims for elk, and for fallow deer, as at Tursac,[57] all Palaeolithic depictions are either of red deer or of reindeer.

In red deer, only the males have antlers, so that these can denote sex, although some figures taken to be hinds may actually be males whose antlers have been shed. In the reindeer, however, both sexes have antlers.

Many depictions of reindeer show clearly the characteristic shape and markings of the species, as in the series of engravings at Trois Frères; ironically, some of the best figures are to be found in Cantabria (Altxerri (Fig.87), Tito Bustillo (Fig.29), Las Monedas), a region where this animal was never plentiful, according to faunal remains.[58]

Occasionally, it is difficult to differentiate between the two types of deer; on the famous baton from Lortet with a scene of deer and fish engraved around it, some scholars see stags, but others see reindeer. (Indeed, Piette first saw them as reindeer, and later as stags!) In the portable engravings of La Colombière, most of the animals thought to be red deer have recently been reinterpreted as reindeer.[59] Similarly, all but one of the 91 deer on the walls of Lascaux have been assumed to be red deer by most specialists of Palaeolithic art;[60] but a number of experts on caribou instead see some of them as reindeer – particularly the series of heads, often called the 'swimming stags', since groups of swimming stags are very rare, whereas swimming caribou are a common sight in the migration season (Fig.84). This alternative identification may well be correct since 74 of the Lascaux cervids (81%) have antlers, and the wide variety of shapes includes some which closely resemble those of tundra caribou and depictions of them by Eskimo hunters during prehistoric and historic times.[61]

As usual, some authors have chosen to see these Palaeolithic figures as accurate depictions of individual animals; thus, the remarkable examples in caves such as Lascaux were detailed records of prize stags (mostly at least 15 years old) with fine antlers,[62] whereas it is far more reasonable to assume that the style and ability of individual artists played a major role, and that the antlers are merely an artistic translation of reality.[63] Attempts

Fig. 84 *The so-called 'swimming' deer of Lascaux (Dordogne). Probably Magdalenian. This panel is c 5 m wide. (JV)*

to differentiate tundra and forest reindeer types from antler shape in the art have also been made, but are unlikely to be reliable.[64]

Pending publication of a detailed anatomical study of deer depictions,[65] one can still make a number of observations: many features, such as the antlers, tail, hoofs, ears and eyes, are often incomplete or done badly – yet, occasonally, even the gland below the eye of the rutting reindeer is depicted, as on the engraving of Kesslerloch, and on seven or eight of the Lascaux deer.[66]

A number of different positions are known for cervids, such as the male bending to lick a female's head at Font de Gaume, the rutting male of Kesslerloch, and other variations at Limeuil and elsewhere – grazing, bellowing, and so forth – and it has been pointed out that both reindeer and red deer are most often depicted in what appears to be the rutting season (autumn/early winter), characterised by features such as the white mane on the reindeer bulls and the thick mane on mature stags, and bodies rounded with fat:[67] for example, a stag from abri Morin has been described as having its antlers thrown back, its ears down, nostrils dilated, mouth open and neck extended.[68]

Ibex

One should really use the term 'ovicaprids' here, since it is quite possible that some Palaeolithic depictions represent the mouflon or even the Tahr *Hemitragus*, and a few have been attributed to these species;[69] but by and large, animals of this type are automatically assumed to be ibex.

The question then arises whether they are the Alpine variety (*Capra ibex*) or the Pyrenean (*Capra pyrenaica*); the former has curvilinear horns, while the latter's horns are more lyre-shaped, and rise at the tips. However, they are often difficult to differentiate in art, for the simple reason that in profile one cannot assess how sinuous or divergent the horns are – indeed, this may be another role for twisted perspective – and in

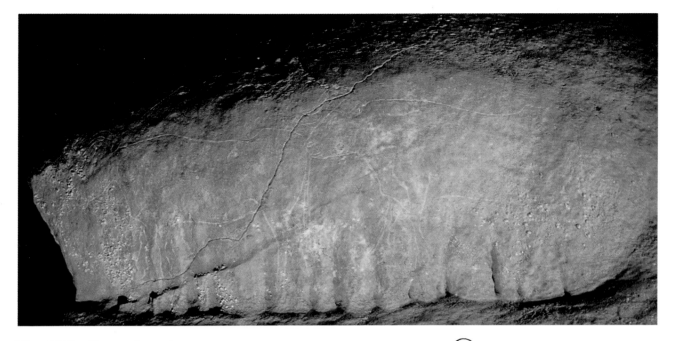

Figs. 85/86   *Engraved aurochs cow (left) and bull (right) from Teyjat (Dordogne), with tracing by Breuil. The animals are 55 and 50 cm long. Magdalenian. (JV). The detail (left) shows the cow's head. (JV)*

Fig. 87   *Engraved reindeer and fox at Altxerri (Guipúzcoa). Magdalenian style. The fox is 25 cm long. (JV)*

Fig. 89   *Photomontage of the decorated panels at Cougnac (Lot): note the use of natural rock formations for the neck and chest of the giant deer. The female ibex at the top (facing left) is 50 cm long; the ibex on the right (and facing right) is 81 cm in length. (JV)*

Fig. 90   *The two headless bears of Ekain (Guipúzcoa). Probably Magdalenian. Panel size: 1.2 m (JV)*

Fig. 91   *Mammoth sculptured in bas-relief in the Grotte du Mammouth (or de Saint-Front), Dordogne. It is 1.25 m high, 1.1 m long, and located 4 m above the present cave floor. (Photo B. & G. Delluc)*

mouth is closed, whereas in these figures, if the jaws were to be closed, they would be in contact with the upper, or even stand behind them.[85] Carnivore canines clearly impressed Palaeolithic people, who, as we have seen, often used them for jewellery; they are the only teeth ever drawn in carnivore figures, and their size is often exaggerated, like deer antlers or bison humps – in one case, however, the large teeth on a figure from Isturitz have been taken at face value as a depiction of the scimitar cat.[86]

Nevertheless, a few felines are accurately portrayed, including those of La Colombière and Labouiche; the 15 of La Marche are fairly realistic – but only 43% of them face right (as opposed to 100% of the bears). We cannot tell for certain whether or not the male big cats had manes, though one or two depictions in Palaeolithic art seem to have one.[87]

The rare mammals

Finally, there is a whole range of animals which are depicted extremely rarely, although often well.[88] Many of them belong to the Magdalenian, and their existence may thus simply be a result of the overall increase in numbers of figures in this period: one can mention the rhino, musk-ox, ass, saïga, chamois, wolf and fox, hare, glutton, otter and possibly the hyena; there is also the remarkable weasel-like animal of the Réseau Clastres (Fig.2). The 'boars' on the Altamira ceiling are now generally thought to be 'streamlined' bison since at least one of them has horns! So far, animals such as small rodents seem to be absent from Palaeolithic art.

Where sea-mammals are concerned, most are seals of various species,[89] some of the portable depictions found hundreds of kilometres inland. Among the finest are the male and female grey seals (*Halchoerus grypus*) on the famous baton of Montgaudier (Charente); they are characterised by their elongated muzzle, their size difference and the male's neck folds. Three enigmatic small shapes above them were recently claimed to be whales seen in the distance.[90] Some parietal figures at Nerja (Málaga) have been interpreted as dolphins, but others believe them to be seals.[91]

Fish, reptiles, birds, insects and plants

Fish are quite plentiful, particularly in the form of portable depictions – unfortunately, Breuil's corpus also includes a mass of decorative motifs which he believed to have been derived from schematised fish; some of them are quite plausible, many others are not.[92] Fish species are especially difficult to identify in Palaeolithic art: most of them seem to be salmonids – sometimes males with a kipe on their lower jaw in the characteristic state of exhaustion after spawning, and thus perhaps representing the autumn.[93]

Marine fish are also represented, both on portable objects (such as the Lespugue flatfish figure, in the central Pyrenees) and on the walls of caves, particularly those near the sea, most notably at Pileta.[94]

One or two tortoises (St Cirq, Marsoulas) and perhaps a turtle (Mas d'Azil) occur in portable art, and one possible salamander from Laugerie-Basse; there are also a number of snakes, although once again Breuil's attributions are excessive, with all manner of wavy lines being interpreted as these creatures – one wonders why eels or worms were not envisaged: the specimens next to the seals on the Montgaudier baton are clearly males.

Birds occur in both types of art;[95] the most recent survey of French specimens found 121 depictions in 46 sites, and once the doubtful cases had been weeded out there remained 81 birds in 31 sites[96] – remarkably few compared to numbers of animal figures. Portable depictions are more abundant than parietal (the latter comprise 15 figures, or 18.5%). Most are Magdalenian (86.4%; 69% are from the mid- and upper Magdalenian alone), while the oldest are all parietal. Water-birds such as swans, geese, ducks and herons seem to be the most numerous (about 37%). The species and even the genus can be very difficult to identify, as the depictions are often mediocre: indeed, in some cases it is not sure whether faces on cave-walls belong to owl-like birds or to humanoids and 'phantoms', although Leroi-Gourhan's belief that the owls of Trois Frères are 'anthropomorphs' seems to rest on his theories about cave topography (see below, p.172) rather than on resemblance.

Another feature which birds can share with humans is a vertical, bipedal posture; recently, portable engravings from Les Eyzies and from Raymonden, traditionally seen as groups of tiny humanoids approaching a bison (in the former) and standing around a bison carcass (in the latter), have been reinterpreted by a specialist on birds as lines of swallows:[97] certainly, their stance is identical, and microscopic study revealed tiny incisions by the heads which seem to be beaks; another piece from Raymonden seems to include two swooping birds.

Only about a dozen definite or probable depictions of insects are known, all in portable art, and mostly Magdalenian.[98] Apart from what seems to be the larva of a reindeer-botfly from Kleine Scheuer, and the famous grasshopper of Enlène, most are beetle-like bugs. However, even the grasshopper, the clearest and most detailed depiction, cannot safely be identified to the genus level.

Finally, plants are also rare in Palaeolithic art,[99] especially in parietal art, and only a very few, such as those on a baton from Veyrier and a pebble from Gourdan, can be classed as definite rather than possible. None can be identified accurately.

To sum up, therefore: the 'animal' category in Palaeolithic art is of limited information value to zoology, and vice versa. The greatest caution and rigorous objectivity must be applied to any attempt at identification of the genera or species, and assessment of other external features and their posture.

It is clear that large herbivores dominate the art, and that other types of creature are either comparatively rare (carnivores, fish, birds, etc), extremely rare (reptiles, insects, plants) or totally absent (rodents, and many other species of mammal and birds). Since Palaeolithic people were certainly familiar with all aspects of their environment, and since birds, fish and plants were important resources for them, at least in the Magdalenian, it follows that Palaeolithic art is not a simple bestiary, not a random accumulation of artistic observations of nature. It has meaning and structure.

This becomes even more obvious when one looks at the differing frequencies of species in a variety of contexts. It has already been mentioned (see above, p.56) that the Tuc d'Audoubert has different species in its portable and its parietal art, and this observation has been repeated elsewhere. Overall, reindeer, fish, birds and plants are primarily found in portable art rather than parietal,[100] while insects are limited to portable depictions.

Some themes are associated with particular forms of artefact: the horse is rare on harpoons and 'baguettes demi-rondes', while bison are not found on spearpoints, are rare on perforated batons, but predominate on plaques.[101] Horses and fish are common on perforated batons.

Different species also predominate in different periods and regions: for example, the bison tends to dominate on cave-walls in Cantabria, although it is rare in the region's portable art, where hinds are predominant on shoulder-blades. The aurochs is more important than the bison in the art of the early phases, but the situation is reversed in the middle and late Magdalenian. The mammoth is of importance in the Rhône and Quercy regions only during the early phases, but in Dordogne during the Magdalenian, when the bison dominates in the Pyrenees.[102] The horse and ibex seem to maintain roughly the same frequency in most places and periods, though the horse appears particularly important in a cluster of Dordogne caves in the Magdalenian (Cap Blanc, Comarque, St Cirq, etc)[103] like the ibex in the late Magdalenian of Cantabria. Such regional and

chronological differences may be at least partly ecological in origin, rather than purely cultural or a matter of 'preference'.

Leroi-Gourhan's calculations concerning species frequency in different regions have recently been presented in the form of graphs; and cumulative percentage frequency graphs have also been produced to compare the depictions in different sites; exercises of this type obviously have to assume that all the figures have been correctly identified, but cumulative percentages are a particularly risky statistic.[104]

Finally, it should be noted that different species are not scattered at random through the sites; this is well known for parietal art (see below,

Fig. 92   *The owls engraved at Trois Frères (Ariège) [see also Fig. 56]. Panel width: 87 cm. (JV, collection Bégouën)*

p.167), but the same phenomenon occurs with portable depictions. The differential distribution of depicted species at Dolní Věstonice has already been mentioned (p.83); at La Madeleine, the horse and bison dominate on the objects of bone and antler, but the reindeer was predominant on a cluster of engraved limestone slabs found in part of the shelter.[105] An excellent example of this phenomenon occurs at Gönnersdorf, where most of the mammoth engravings were found inside a habitation near the hearth, with another concentration a few metres away; depictions of birds, on the other hand, were concentrated precisely in those zones devoid of mammoths, whereas horses and humans were found all over the excavated area.[106] Since the mammoth engravings came from a winter habitation (according to the faunal remains) and the birds from a summer one, it is possible that there is a link between season and depicted species at this site.

Humans

The category of humans has traditionally been given the unwieldy title of 'anthropomorphs', comprising not only unmistakable human forms but also images which overlap with the other two categories, and many which seem to have been included simply because they looked figurative but did not resemble any known animal. The fact that humans lack the specialised anatomical features of other mammals (horns, antlers, etc) meant that all kinds of vague, amorphous images were unjustifiably lumped together with clear depictions of people.[107] Instead, a distinction should be made between definite humans, 'humanoids' and composites, and the rest should be left undetermined.

Definite humans

Perhaps the clearest images in this group are the painted stencils and prints of hands which, as we have seen (p.104), are quite numerous and can dominate the art of whole caves (eg Fuente del Salín) or parts of caves (Gargas, Castillo). Apart from a stencil on a block from abri Labattut (which may have fallen from the wall), found between two Gravettian layers, all hands are parietal; only the Labattut hand has any sort of date, but studies of superposition, composition and association in different caves suggest that this simple motif spans the entire Upper Palaeolithic.

As mentioned earlier, a few caves have incomplete hands, which may be due to mutilation, pathologies, or some system of signals and gestures using bent fingers. Ethnography provides a wide range of possible explanations for hand stencils, both intact and incomplete: eg a signature, a property mark, a memorial, a wish to leave a mark in some sacred place, a record of growth, 'I was here', or simply 'just put there'. In view of the timespan involved, it is probable that all these roles and many more were involved in the Palaeolithic stencils; and since the specific reasons for making such marks seem to be remembered for only a couple of generations[108] it is unlikely that we shall ever be certain why Palaeolithic people painted them.

For decades, despite the steadily growing number of 'Venus figurines', it was believed that depictions of humans were rare in Palaeolithic art, and that they were badly done (even though many animal figures were equally incomplete or sketchy!). Now that at least 115 quite realistic human figures are known among the engravings of La Marche,[109] outnumbering all other species depicted at the site, it is clear that they are not so rare; their quality, together with that of the figures at nearby Angles-sur-l'Anglin, and that of the ivory heads from Brassempouy and Dolni Věstonice, shows that realistic images of people were by no means taboo, as had been supposed.

Nevertheless, it remains true that few depictions of humans can match the finest animal images in detail and beauty. More effort may have been put into the animal figures for some ritual purpose, but the answer might simply be lack of ability. Drawings of humans and animals require different skills, and it is noteworthy that Breuil, who learned much of his art from the 'Palaeolithic school', had a talent that was 'confined to portraits of animals. When he ventured to reproduce human figures … his work was not on a higher level than that of any fairly efficient amateur.'[110] Breuil drew amusing caricatures of his teachers at school, and it is quite possible that some of the Palaeolithic portraits which look funny to our eyes are actually caricatures rather than attempts at serious portraits.

Genitalia are rarely depicted on Palaeolithic humans, even on figurines.[111] On bodies drawn in profile, the phallus can be seen, but the vulva cannot. Where genitals are absent, or on isolated heads, males can still be confidently recognised from beards and moustaches, whereas breasts denote females (though it should be noted that the Brno male statuette has small breasts or exaggerated nipples). At La Marche, using these criteria, there are 13 definite males (including 11 with beards).[112]

Of 51 bodies drawn at the site, 4 are definitely male, but only 27 of the rest can confidently be seen as female: 8 of these have breasts; the others have been identified on the basis of the size of hips and buttocks.[113]

Heads without facial hair have to be left 'neutral' – length of hair is not a sure guide. Hence the Brassempouy head is unsexed, although it is usually called a 'Venus' or 'Lady', and it has even been placed on a female body in

Fig. 93 *Tracing of engraved plaquette from La Marche (Vienne), and the face of a bearded man extracted from the mass. Magdalenian. (After Airvaux & Pradel)*

0   1   2cm

one reconstruction.[114] Other scholars have attempted to identify males on the basis of nose-shape or faces that jut forward[115] – despite the fact that the two females engraved at Isturitz have equally jutting faces (Fig.100). There is a wide range of nose- and face-shapes at La Marche alone.

The presence of adornments can denote females, though it is not an infallible guide: the nine bodies at La Marche that seem to have bracelets and anklets are all female; analogous ornaments including 'necklaces' can be seen on the above-mentioned females from Isturitz (identified by their breasts), as well as on the similar *Femme au Renne* from Laugerie-Basse (who has a vulva but no breasts), and on a number of female figurines from the Soviet Union.

Clothing is rarely clear; the 'Venus of Lespugue' (Fig.94) seems to have a garment of some kind showing at the back, and belts are depicted occasionally (eg at La Marche and Kostenki). The bearded male, sculptured, engraved and painted at Angles-sur-l'Anglin, appears to be wearing a garment with a fur collar, and perhaps also some form of headgear.[116] However, the so-called *Femme à l'anorak* at Gabillou is unsexed, and the 'hood' may simply be a hair-line. Elaborate hairstyling is extremely rare (eg the Brassempouy head, the 'Venus' of Willendorf).

Using the strict criteria outlined above, it has been established that La Marche has 13 male, 27 female, and 69 undetermined humans; a study of 410 parietal and portable human figures from western Europe (including those of La Marche but not those of Gönnersdorf) produced a very similar picture: approximately 10% male, 25% female and 65% neutral. However, 70 figures from eastern Europe – all in portable art – proved to be approximately 4% male, 60% female and only 36% neutral.[117] The difference with the west is largely due to the fact that almost all the eastern figures are statuettes, on which gender tends to be more easily recognisable.

In Palaeolithic depictions of humans as a whole, details such as eyebrows, nostrils, navels and nipples are extremely rare. The legs are often too short (as in the art of many later cultures); the legs and/or feet of figurines are held together or slightly apart. Few figures have hands or fingers drawn in any detail (examples include La Marche, the 'Venuses' of Laussel and Willendorf, the man in the Lascaux shaft-scene, etc). Arms on statuettes are usually held close to the body, for technical reasons, and rest on the stomach or, sometimes, the breasts. On other figures, the arms may be at the sides, raised horizontally (as on the Sous-Grand-Lac man), or up in the air – all these poses are represented at La Marche.[118]

Except for the two ivory heads mentioned earlier, very few statuettes (apart from those of Siberia) have any kind of facial detail; their heads are held erect or tilt forward slightly (though one from Kostenki tilts upwards). Many Palaeolithic humans are headless: in some cases, as in the female bas-reliefs of Angles-sur-l'Anglin, it is clear that they never had heads. The same is true of nine engraved humans at La Marche, but others at this site and elsewhere may have had their heads broken off, either purposely or accidentally.[119]

There are three heads seen full-face at La Marche; all the others, together with 90% of the bodies, are in profile, and it is therefore worth examining which way they face. Of 57 isolated heads, 35 face right and 22 left; of 51 bodies, 33 face right and 18 left. In short, of 108 people, 68 (63%) face right and 40 (37%) left. Once again, as with sexing (see above), the percentages are found to be consistent when the analysis is extended to 167 humans from other sites: 100 (60%) of them face right, and 67 (40%) left.[120]

The depiction of women

Apart from a few definite males (eg Brno, Hohlenstein-Stadel) (Fig.49), most carvings of humans are female. The 'Venus' figurines[121] have been presented in so many art histories and popular works on prehistory that they have come to characterise the period and its depiction of women; this is unfortunate, partly because such statuettes are rare when seen against the timescale of 25 millennia, and partly because the constant display of a few specimens with extreme proportions presents a distorted view.

The term 'Venus' was first used by the Marquis de Vibraye in the 1860s in connection with his '*Vénus impudique*' from Laugerie-Basse (ironically, a very slim lady!), and was later adopted by Piette for the more corpulent figures from Brassempouy and elsewhere; subsequently it seems to have become attached primarily to the obese statuettes.[122]

Despite the accepted view that Palaeolithic depictions of humans were badly done, early scholars nevertheless tried to use them as evidence of different races during the period. As with similar attempts involving horses and other species (see above), they were doomed to failure. Piette was among the first to have a try; apart from his erroneous belief that a Grimaldi carved head was negroid, his major mistake was to see the obese 'Venuses' as steatopygous: ie having the special fatty deposits which produce the massive, high and wide buttocks of some female Bushmen and Hottentots.[123]

In fact, he was confusing this phenomenon with the kind of buttock development which is found in all races. Only one statuette (the 'Polichinelle' of Grimaldi) could conceivably be steatopygous; the rest merely present proportions which one can see at any time and anywhere on women who have produced lots of children. In short, 'Venus' figurines have no value whatsoever as indicators of race.[124] Only four (Lespugue, and one each from Willendorf, Grimaldi and Gagarino) have really extreme proportions, and only Lespugue has truly monumental breasts (Fig.94). On the whole, the obese carvings are not anatomically abnormal – they are simply bodies worn and altered by age and childbearing. They seem well nourished, with their adipose tissue concentrated in discrete areas rather than spread out as a continuous layer.

One recent study has looked at 132 Palaeolithic 'Venuses' and has tried to estimate their age group, dividing them into the following categories: young (pre-reproductive, with a firm body, high breasts and a flat stomach); middle-aged (reproductive and potentially pregnant, with a fleshy body, big breasts and a protruding stomach); and old (post-reproductive, sagging all over). The result was 30 (22.7%) young, 23 (17.4%) middle-aged and apparently pregnant, 50 (37.9%) middle-aged but not pregnant, and 29 (22%) old.[125] Despite its tentative and subjective nature, this exercise does at least suggest that the carvings probably represent women throughout their adult life.

As with bison and other species, some scholars have attempted to establish the measurements and conventions used in the construction of Palaeolithic human figures, but made the same mistake of using the images themselves, and not human anatomy, as the base of the study. Drawing lozenges around the Venuses merely shows that they have the same basic shape (as one would expect of human depictions), though with variations, and exaggerates the degree of their deformation. Anatomical studies, on the other hand, suggest that the figures' proportions are close to reality.[126]

Fig. 94   *The 'Venus' of Lespugue (Haute Garonne) seen from the side. Ivory, probably Gravettian. Height: 14.7 cm. (JV). (Drawings show front and back views after Pales and Leroi-Gourhan respectively)*

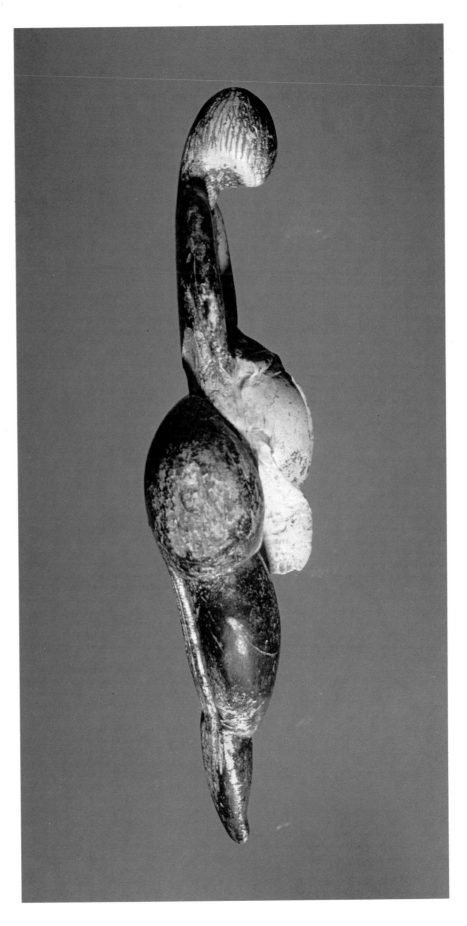

As was mentioned earlier (p.85), 'Venuses' are not limited to the Gravettian and Magdalenian; there is some reason to believe that those of Brassempouy, at least, are Aurignacian; a possible 'rough' for one was reported in the Solutrean of Roc de Sers,[127] but is a most unconvincing Venus. So far, however, only Angles-sur-l'Anglin has produced both parietal (four in bas-relief) and portable specimens (two statuettes), together with over a dozen other human depictions, including the famous large and small 'versions' of a male 'portrait'.[128] However, the 'Venus' figurines were not found in any kind of privileged position. Very few such statuettes in western Europe have any stratigraphic provenance, though some (eg Lespugue, Courbet) seem to have been carefully hidden under rocks in caves and shelters. In eastern Europe, however, the excavation of open-air habitations has frequently encountered intact female figurines in special pits in hut-floors:[129] for example, one from Kostenki I, found in 1983 and dating to about 23,000 years ago, was upright in a small pit, leaning against the wall and facing the centre of the living area and the hearths; the pit was filled with soil mixed with red ochre and was capped by a mammoth shoulder-blade.[130] The eastern statuettes are often interpreted by Soviet scholars as a mother- or ancestor-figure, a mistress of the house.

These hidden figurines, in both western and eastern Europe, cast doubt on the theory that Venuses were meant for public exhibition, and that those with a protuberance at the base (Tursac, Sireuil) or with peg-like feet were hafted on to stakes or stuck in the ground for display.[131]

Finally, it is worth noting that no definite 'Venuses', either parietal or portable, have yet been found in the Iberian Peninsula; this is partly because bas-reliefs are limited to parts of France; but in view of the remarkable distribution of female statuettes from the western Pyrenees to Siberia, there seems no reason why Spain and Portugal should not contain a few, and no doubt future discoveries will alter the situation.

'Humanoids'

This category should comprise all those figures interpreted, but not positively identified, as being human. For example, there are many heads which could belong to animals but also resemble some of the more 'bestialised' heads on definite human figures: one, engraved on the wall at Comarque, represents a bear for some scholars, but others see it as human.[132] It cannot safely be assigned to either group. A number of rudimentary and sometimes grotesque heads on cave-walls, seen full-face, are potentially human (eg Marsoulas, or Trois Frères), while others (such as 'phantoms') are best left undetermined. As mentioned above (pp.132-3), the miniature figures grouped on some engraved bones have traditionally been interpreted as human (despite the lack of any distinguishing features) but may in fact be realistic birds.

This is by no means a new problem. From the start, Breuil was well aware of the difficulties in separating 'elementary anthropomorphs' not only from animals but also from 'signs'. It occurred to him that the 'claviform' (clublike) signs might be derived from stylised females, but he dropped the idea because too many intermediate steps in the stylisation were missing.[133] Today, however, we have far more 'stylised females', and the idea has re-emerged.

Such stylisations are particularly abundant in the late Magdalenian, though they are not exclusive to the period. They occur both on cave walls

Fig. 95 *Stylised humans engraved on plaquette from Gönnersdorf (Germany): two female figures are depicted, 'face to face'. Magdalenian. The figure on the right is 11.3 cm long. (Photo G. Bosinski)*

Fig. 96 *Engraved plaquette from Gönnersdorf (Germany), showing four schematised females with what may be a small child between them. Magdalenian. Width of plaquette: 8 cm. (Photo G. Bosinski)*

(eg Fronsac[134]) and on slabs and plaquettes as at Gare de Couze and La Roche-Lalinde: Gönnersdorf alone has 224 such figures on 87 engraved surfaces, as well as about a dozen figurines of similar type, in ivory, antler and schist; some of the figurines are perforated.[135]

The engraved figures are all stereotyped headless profiles (mostly facing right) with protruding buttocks; about a third can safely be identified as female humans because they appear to have breasts, as well as arms and other details. The other, more sketchy outlines from this site and elsewhere have been interpreted as females because of their resemblance to the more definite examples,[136] but they could be either young females or males (none is definitely male), and are best left as 'neutral': some of them would probably be classed as non-figurative if found out of context!

The Gönnersdorf females have a wide range of proportions, and occur both singly and in groups of two, three, and more, which have been interpreted as 'dances' for some reason. Some couples face each other, and one woman has what may be a baby on her back (Figs.95-6).[137]

In view of the position of Venus statuettes in open-air habitations (see above), it is important to note that the intact figurines from Gönnersdorf were all found in pits.[138] It will be recalled that such statuettes are thought to have been made for long-term use while plaquettes had a short-term use before being discarded and/or broken (see pp.54 and 77).

Various authors have proposed unilinear sequences of development from realistic to stylised, and even to 'signs'; Breuil did this for a number of species, and Leroi-Gourhan saw humans as being stylised into the headless profiles and ultimately into 'claviform' signs which, like the profiles, occur singly and in groups.[139] Alas, all such schemes are to some degree subjective, and often ignore dating and other factors: for example, true claviform signs are exclusively parietal, probably mid-Magdalenian in date, and limited to the Pyrenees and Cantabria (and possibly also earlier at Lascaux); but the headless female profiles are most abundant in Germany (and to a lesser extent in Dordogne), and seem to be late Magdalenian! The claviform comprises a vertical line with a bulge at one side; but unlike the buttocks on the stylised females, the bulge is usually on the middle or the upper half of the 'sign'![140] It has been suggested that the 'upper bulge' on a claviform instead represents breasts, but this is based on supposedly stylised statuettes from Dolní Věstonice and elsewhere, comprising a rod with two little lumps near the top, and thus on analogies with sites even further away in space and time from the claviforms. In any case, stylisation of humans or of any other theme can and probably did take place independently, in a number of phases and regions.[141]

The claviform question demonstrates the dangers of subjective impressions in these matters; a similar debate concerns some carvings from Mezin (Soviet Union) which different scholars see as schematic images of women, as phallic symbols or as birds.[142] They may, in fact, be schematic female symbols (triangles, etc) engraved on phallic birds! Clearly, in situations of this kind, the figures must be left undetermined, and speculation about what they depict must never be claimed as a firm identification.

Composites

We have just encountered some undiagnostic examples which could be either animal or human; but what of those figures which have clear and detailed elements of both? They been called 'anthropozoomorphs' or 'therianthropes', but the term 'composites' is simpler, and does not give priority to either the human or the animal features.

In the past, thanks to the dominance of the 'hunting magic' theories (see below, p.150), all such figures were automatically and unjustifiably called 'sorcerers', and were assumed to be a shaman or medicine man in a mask or an animal costume.

But how can we differentiate between people wearing masks, people with jutting, bestialised faces (see above) and humans with animal heads? Unfortunately we can't, and it is likely that all the above are represented in Palaeolithic images. Inevitably, interpretations have been heavily influenced by ethnographic parallels: for example, a bison-headed humanoid at Trois Frères (with the supposed 'musical bow') (see Fig.104) bears some resemblance to North American Indians disguised as bison for a dance,[143] while an equally famous figure from the same cave – the 'Sorcerer', with his antlers and other animal parts (Fig.97) – looks just like an eighteenth-century depiction of a Siberian shaman.[144] These

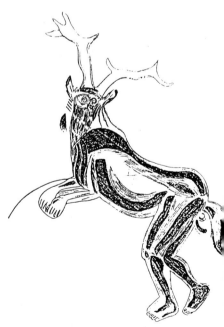

Fig. 97 *Photomontage of the 'sorcerer' of Trois Frères (Ariège), together with Breuil's version of this painted and engraved figure. Probably Magdalenian. Length: 75 cm. (JV, collection Bégouën)*

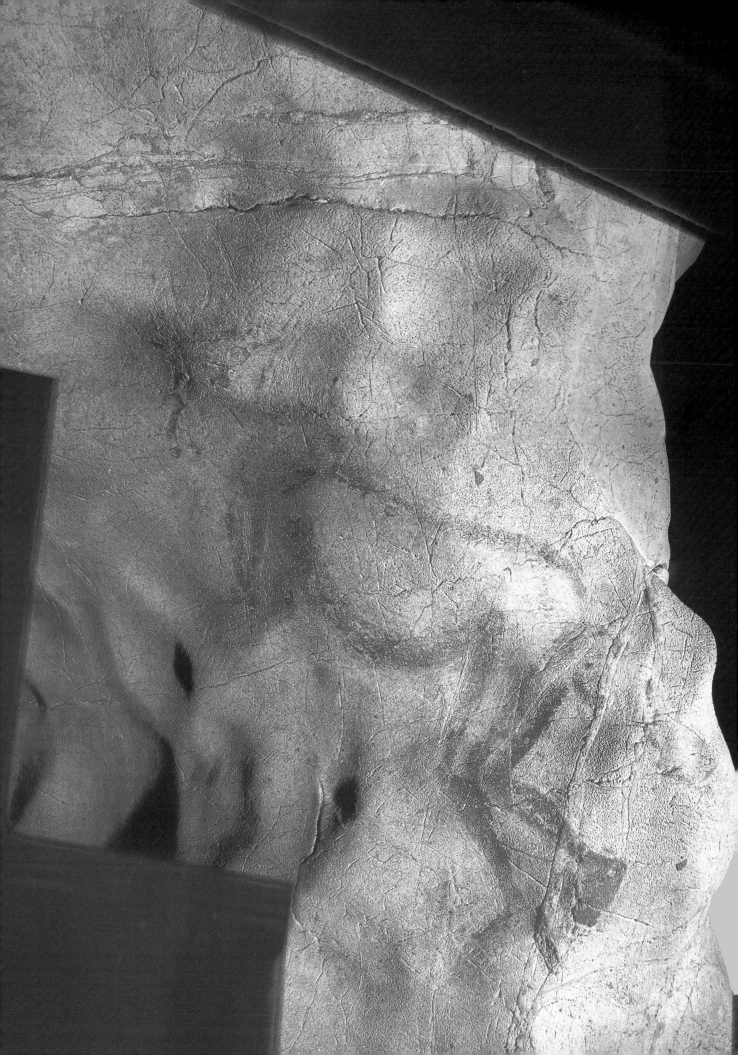

Palaeolithic images may indeed be disguised humans, although it cannot be proved.

On the other hand, many scholars have noticed the strange resemblance between the Hohlenstein-Stadel statuette (Fig.49) and Egyptian figures of a deity with a human body topped by a lion's head; in this case, therefore, the ethnographic analogy points towards an imaginary being – again without proof. Likewise, Breuil called the Trois Frères 'sorcerer' the 'God of the Cave' and considered him an imaginary figure, dominating the 'sanctuary'. Not all such 'sorcerer' figures, however, are in imposing positions – indeed, some are hidden away among other images, and can be very hard to see.[145]

The Trois Frères figure is certainly strange: only the upright position and the legs and hands are really human. The rest is a mixture of different animals – the back and ears of a herbivore, the antlers of a reindeer, the tail of a horse, and the phallus in the position a feline would have it.[146] Very few such composite figures have horns or antlers (apart from the Trois Frères examples, Gabillou has the clearest); overall, composites or 'monsters' are fairly rare in parietal art – according to Leroi-Gourhan, only about 15 sites have them, with no more than half a dozen in each, and some of those he includes are probably not even humanoid.[147]

The distribution of human figures is interesting: as with parietal and portable art, there are many sites with none, some with one or two, while others like La Marche or Gönnersdorf have many. Pales' sample of 410 in western Europe (including La Marche but excluding Gönnersdorf) comprised 187 realistic figures, 214 humanoids, and 9 composites[148] – the vast majority (c 80%) were engravings, with carvings accounting for almost all the others: painted humans are extremely rare.

Definite humans are scarce in parietal art, and there are few women resembling the Venus figurines or the stylised profiles (examples include Angles-sur-l'Anglin, Laussel, Fronsac). The distribution of sexed and unsexed humans within caves will be studied later (p.172).

Portable art accounts for over 75% of Palaeolithic human depictions; as mentioned earlier, at Dolní Věstonice the human figures were not found in the same hut as those of herbivores, but were with carnivore figures. At Gönnersdorf they were scattered throughout the excavated area, unlike engravings of mammoths and birds: animals predominated on some slates at this site, while other slates were primarily devoted to stylised females; no intentional grouping of human figures with animals or with symbols could be discerned.[149]

Non-figurative: 'signs'

The non-figurative category of Palaeolithic markings was neglected until relatively recently, for the simple reason that it seemed uninteresting, or impossible to explain or define. Many lines, as mentioned earlier (p.44), were simply ignored as 'traits parasites'. Nowadays, however, thanks to the attention paid to the abstract 'signs' by Leroi-Gourhan and to the discovery of similar non-figurative motifs in Australia and elsewhere (see above, p.28), we have to come to terms with the possibility that these marks may have been of equal, if not greater, importance to Palaeolithic people than the 'recognisable' figures to which we have devoted so much attention. Certainly, it has been estimated that non-figurative marks are two or three times more abundant than figurative, and in some areas far more: for example, on 1,200 engraved pieces of bone and antler from 26

Magdalenian sites in Cantabria, there are only 70 identifiable animals – all other motifs seem non-figurative.[150]

As already mentioned, it is hard to separate this category from the two others; and it comprises a tremendously wide range of motifs, from a single dot or line to complex constructions, and to extensive panels of apparently unstructured linear marks. There are significant differences in content between portable and parietal examples, which make it extremely difficult to date parietal 'signs', particularly in cases where they constitute the only decoration and figurative images are absent, as at Santián and other sites.[151] Even in caves with both figurative and non-figurative images on the walls, 'signs' can be either totally isolated, clustered on their own panels or in their own chambers, or closely associated with the figurative – and sometimes all of these.

In the past, certain shapes were assumed to be narrative or pictographic: ie to represent schematised objects on the basis of ethnographic comparisons (see Chapter 7, below) or, more often, subjective assessment of what they 'looked like': hence, we have terms such as 'tectiforms' (huts), 'claviforms' (clubs), 'scutiforms' (shields), 'aviforms' (birds), 'penniforms' (feathers), etc. These are no longer taken literally but, as they have entered the literature so decisively, are retained simply as rough guides to particular shapes. Even if these motifs were ultimately derived from real objects, they could now be purely abstract,[152] although it is impossible for us to separate them fully from 'realistic' figures: some signs may indeed be schematised or abbreviated (or unfinished!) versions of some object. Some authors now regard 'signs' as 'ideomorphs' – ie representations of ideas rather than objects, but we simply don't know whether they are real or abstract or both.

We inevitably apply our understanding to what we see; our background, culture and art history come into play, and hence the early scholars tried to translate some signs as if they were hieroglyphics, in the vain hope of finding a Palaeolithic Rosetta Stone. Even if these signs *were* representational, their meaning would be far from straightforward without an informant: for example, among the Walbiri of Australia, a simple circle can mean a hill, tree, campsite, circular path, egg, breast, nipple, entrance into or exit from the ground and a host of other things; similarly, 'translation' of simple signs and pictographs in North America can often produce differing and unexpected results.[153] A mere resemblance of shape does not necessarily mean that a motif is an image of an object.

What is clear is that the simpler motifs are more abundant and widespread, as one would expect, since they could have been invented in many places and periods independently. The more complex forms, however, show extraordinary variability and are more restricted in space and time, to the extent that they have been seen as 'ethnic markers', perhaps delineating Palaeolithic groups of some sort:[154] hence, 'quadrilateral' signs seem concentrated in Dordogne, particularly at Lascaux and Gabillou; a different kind of 'quadrilateral' is found in a Cantabrian group of caves (Castillo, La Pasiega and Las Chimeneas, all in the same hill, together with Altamira, some 20 km away); 'tectiforms' are found in the Les Eyzies region (Dordogne), in a small group of caves only a few kilometres apart, including Font de Gaume and Bernifal;[155] very similar 'aviforms' are known at Cougnac and Pech Merle, 35 km apart; triangles are predominantly found in the centre of Spain, but are rare in Cantabria; perhaps most remarkable of all is the distribution of true 'claviforms', which are known in Ariège (Niaux, Trois Frères, Tuc d'Audoubert, Le Portel, Fontanet, Mas d'Azil) and in Cantabria (Pindal, La Cullalvera) some 500 km away (and perhaps also at Lascaux).

As will be seen below (p.169), Leroi-Gourhan divided 'signs' into two basic groups, the 'thin' and the 'full', which he linked to a sexual symbolism (phallic and vulvar); he later added a third group, that of the dots.[156] In view of the wide range of shapes on display, such a division was far too simplistic and schematic; using geometric criteria, other scholars have proposed seven or even a dozen classes of parietal sign but still found some very hard to fit in.[157]

Similarly, the non-figurative motifs in portable art have been receiving an increasing amount of attention lately, and likewise have been divided into different groups.[158] Meg Conkey, for example, has sought out the basic units of decorative systems in Cantabrian portable art, and their structural interrelations. In a sample of 1,200 pieces, she found a set of about 200 distinct 'design elements', which she organised into 57 'classes'.[159] Her analysis revealed that a core set of 15 elements was used throughout the Magdalenian, and was widespread within and outside the region. As with the parietal signs, therefore, it will be necessary to focus on the more complex designs in order to identify regional variations in style (and possibly distinct social groups of some kind).

In both parietal and portable art, a full survey of the presence and interrelationships of different motifs, as well as of their association with each other and with other figures, is required, but will entail a more complete published corpus than we have, followed by computer analysis. It is the presence and absence of particular combinations which is revealing: in parietal signs, for example, very few binary combinations occur out of the range of possibilities, and only signs found in binary combinations also occur in 'triads'.[160] Clearly these marks were not set down at random, but follow some set of rules and simple laws.

Can they therefore be seen as a primitive form of writing? Theories about this go back to the discovery of portable art, when a variety of enigmatic motifs associated with animal figures were seen as possible artists' signatures by a number of scholars including Lartet, Garrigou and Piette.[161] It is inevitable that, despite their wide variety of shapes, some of the Palaeolithic signs will resemble some of the simpler characters in certain early forms of writing; after all, the range of possible basic marks is somewhat limited. It has recently been claimed that some Palaeolithic signs have very close analogies with characters and letters in ancient written languages in the Mediterranean, the Indus valley and China,[162] but, as with resemblances to objects, this does not necessarily prove anything. As far as parietal signs are concerned, it often appears to be their presence which was important, rather than their layout or orientation;[163] instead of forming a script, they were sometimes joined through superposition, juxtaposition or actual integration to form composites.

Where motifs on portable objects are concerned, however, it is still possible that they include some sort of 'pre-writing'; we simply don't know. What is almost certain is that the meaning of the signs and marks, no matter how abstract they may appear to us, must have been clear to the maker and to those who saw and/or used them. We can see this today with our road signs and warnings: some have meanings obvious to everyone, others have to be learnt, but all are known to those who operate within that system.[164]

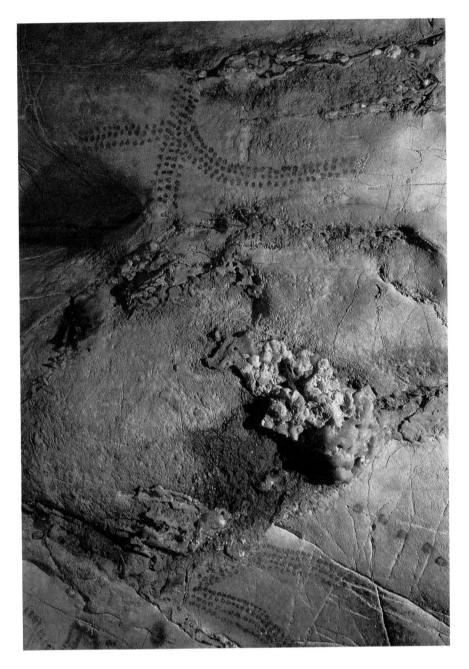

Fig. 98 *'Signs' in the cave of Castillo (Santander). Width: 58 cm. (JV)*

Colour schemes

A final factor in the content of Palaeolithic art which we have yet to consider is the use of different colours; we have already seen that the range was limited and usually involved a straight choice between red and black. At first sight, it might seem possible to present a theory that one of these represented males and the other females (for example in hand stencils), just as we have blue for boys and pink for girls; but the situation is far more complex.

Where signs are concerned, their colour seems linked to different regions and periods, but red usually dominates. As mentioned earlier (p.110), red is more visible in dark caves, so that whatever message (of warning, topographic guidance, instruction, etc) the signs conveyed could readily be seen and understood. In the Iberian Peninsula, 23% of the parietal signs are engraved, 27% painted in black, and no less than 50% painted in red;[165] in France too, about two-thirds of painted signs are red, and they are often the only (or almost the only) red figures in the cave.[166] This is particularly noticeable on friezes of black outline animals, such as Niaux's Salon Noir, or the frieze at Pech Merle (Fig.73).

The use of colour for both signs and animal figures alters significantly through space and time: a survey of 38 caves in France and Spain found that black accounted for 48% of figures and red for 43.7%, only 5.3% in yellow and brown, and mixtures (bichromes and polychromes) a mere 3%[167] – a statistic which underlines the fact that the latter are exceptional and unrepresentative of Palaeolithic art, despite their popularity in art-history books.

However, when the paintings were dated according to Leroi-Gourhan's styles (see above, p.62), a difference was found: in Style III (late Solutrean/ early Magdalenian), 56% of figures were red, and 34.7% black, while in Style IV (mid- and late Magdalenian) these percentages were almost exactly reversed, with black animals becoming particularly dominant at the end of the period. Style IV appears to have a marked tendency for black animals and red signs.

Black and red, whether in signs or figures, are sometimes found in different parts of the same cave (eg Pech Merle, Castillo, Altamira), which may denote a chronological difference, or some purposeful colour scheme. Some signs are associated with a particular colour: about 94% of claviforms are red, as are 100% of painted tectiforms in Dordogne; most dots are red, and isolated dots are always red.[168]

There are also a few such associations discernible in animal figures, but these appear to be of regional rather than general significance: for example, in Spain 95% of hinds are red, whereas stags can be either colour; Las Chimeneas has only stags, and they are all black, whereas all the 17 hinds in Covalanas are red. At Castillo and La Pasiega, however, red dominates for deer, no matter what their sex.[169]

All of the above should have made it clear that in Palaeolithic art there are no absolute rules, no certainties, and a very broad range of complex problems; but also that this is no random accumulation of pictures – on the contrary, it is characterised throughout by careful selection of surface, size, technique, colour, species, anatomical detail, degree of accuracy or stylisation. It is now necessary to examine the sense that different people over the last century have tried to make of all these facts, and in what directions current research is taking us.

7

Reading the Messages

So far, we have looked at Palaeolithic art and asked questions about where, when, how and what; it is now time to grasp the nettle and tackle the most difficult topic of all: why?

Early theories: art for art's sake

The first and simplest theory put forward to explain the existence of art in this period was that it had no meaning; it was simply idle doodlings, graffiti, play activity: mindless decoration by hunters with time on their hands. Such a view, formulated most clearly by Lartet and Piette in the nineteenth century, reflected not only the anticlericalism of de Mortillet (see p.21) who refused to believe that Palaeolithic people had any religion, but also the spirit of the age:[1] art was still seen in terms of the recent centuries, with their portraits, landscapes and narrative pictures. It was simply 'art', its sole function was to please and to decorate. In addition, Palaeolithic art indicated that the people of the period had plenty of leisure time, implying that hunting was easy, there was lots to eat, and life was pleasant, calm and (in Piette's opinion) largely sedentary.

The 'art for art's sake' view arose from the first discoveries of Palaeolithic portable art in the 1860s and 1870s, and it must be admitted that there was little in the content and composition of these pieces to suggest any deeper purpose; indeed, it remains quite possible that some of the decoration of portable objects, and especially of utilitarian items, may simply *be* decoration – although more may be involved even here, from marks of ownership to the complex notations and seasonal images studied by Marshack (see below, p.183). Similarly, the newly discovered engravings in the open air do not seem as enigmatic as those inside caves; but it should be remembered that those of Campôme are all incomplete figures, not simple pictures of animals which happened to be passing.

Of course, when parietal art began to be found, it rapidly became clear that something more was involved: as will be seen later in this chapter, the restricted range of species depicted, their frequent inaccessibility and their associations in caves, the palimpsests and undecorated objects or panels, the enigmatic signs, the many figures which are purposely incomplete, ambiguous or broken, and the caves which were decorated but apparently not inhabited, all combine to suggest that there is complex meaning behind both the subject-matter and the location of Palaeolithic figures. There are patterns that require explanation, repeated patterns which suggest that individual artistic inspiration was subject to some more widespread system of thought. It is worth noting that, even in 1878, when only a few pieces of portable art were known, Cartailhac had already sensed that there was a purpose and a meaning behind this art, having probably been influenced by the writings of Tylor.[2]

Nevertheless, despite all the above evidence, the theory has recently surfaced again,[3] with a claim that the art has no meaning, no religious, mythic, magical or practical purpose; it is simply play, mere aesthetic activity indicative of an inherent human desire to express ourselves artistically. This view resembles the assumption that someone is innocent until proven guilty: *nothing* in the periods before writing can be proved absolutely to be religious; but if Palaeolithic art simply comprised 'objects of contemplation', one would expect to find a pretty broad spectrum of subjects, accumulated more or less randomly on suitable surfaces. Yet the opposite is the case. Besides, countless studies of modern 'primitive' cultures make it clear that there is not necessarily a boundary between the sacred and the profane – *everything* can have a meaning for them, whether it be decoration, dances or mere gestures.[4] To compare Palaeolithic art to modern graffiti, or to attribute it to some inherent human desire to leave a personal mark in places which are isolated or which stir the emotions, is a projection of the present into the past.

It is undeniable that the artists must sometimes have derived great personal satisfaction, and perhaps enhanced prestige, from the best of their work, and certain 'realistic' figures are probably finer and more detailed than was strictly necessary to make their message intelligible – just as medieval craftsmen lovingly put minute details into stained glass which was placed so high up that nobody could see them without binoculars!

However, simple aesthetics explains nothing and has no bearing on meaning. An aesthetic interpretation 'reduces cultural phenomena to an innate tendency and directs explanation inward to mental states about which we can know nothing' – to infer a state of mind (pleasure) from the art, and then use it as an explanation is a circular argument![5] The theory tells us more about its proponents' reaction to Palaeolithic art than about the artists.

Hunting magic, hunting and shamanism

After the death of Piette in 1906, utilitarian theories took over, and primarily that of hunting magic. This was largely caused by the newly published (1899) account by Spencer and Gillen of the life and beliefs of the Arunta of central Australia, which inspired Reinach[6] to equate these 'primitive' users of stone tools with those of the Palaeolithic, and therefore to assume that the same motivation lay behind the art of both cultures.

The Arunta were said to perform ceremonies in order to multiply the numbers of the animals, and for this purpose painted likenesses of these species on rocks. The analogy seemed perfect, and Palaeolithic art must therefore have had a magico-religious cause. For a few decades, all ideas and interpretations were subjectively chosen in this way from the ever-growing mass of ethnographic material from around the world. This is quite understandable – it appeared self-evident at the time that 'primitive' societies provided lots of precious information about the making and function of rock-art in the past. Almost a third of the first monograph on cave-art was therefore devoted to a review of the available ethnographic evidence: the present would teach us about the past, through the parallelism of the artistic activity of all primitive hunting peoples.[7]

'Sympathetic magic', including hunting magic, operates on the same basis as pins in a wax doll: ie the effect resembles the cause. The depictions of animals were produced in order to control or influence the real animals in some way: any injury to the image would likewise be inflicted on its subject. Desirable effects such as ensuring that the prey would be as fine and well fed as the image were also possible. In short, this kind of art is intended to exert influence at a distance, rather like modern advertisements; it is purely functional, and directly related to the need for food.

Breuil adopted the 'hunting magic' view because he felt that Palaeolithic art sprang from the hunters' anxieties about the availability of game. Most other scholars followed, particularly Bégouën who saw evidence for ritual and magic in almost every aspect of Palaeolithic art.[8] One feature which he constantly stressed was that some images seemed to have been 'killed' ritually with images of missiles , or even physically attacked with a weapon or by hand, presumably to improve the chances of the hunt, or (in the case of carnivore figures) to vanquish a fearsome enemy. Marks at the mouths or nostrils of animals were always interpreted as blood being vomited by a dying beast; marks on their bodies were always wounds or missiles, and dots were stones thrown at them: the famous bear of Trois Frères, covered in 189 small circles, was thus an animal stoned to death, with blood pouring out of its mouth.

The theory accounted for the broken and short-lived engravings on plaquettes: once the animal had been 'killed' and the ritual was over, the image had no further use or significance, and was discarded. It also had an explanation for why so many figures were drawn on the same pebble (as at La Colombière) or panel when so many equally good stones or surfaces were available but left blank: they were 'good hunting' pebbles or panels. If the first figure drawn there produced the desired result, they were used again and again.[9]

Such assumptions also influenced or dictated the intepretation of 'signs', which were thought to represent whatever object they looked like: the 'claviforms' were clearly clubs or throwing-sticks, and 'penniforms' were arrows, while the Dordogne 'tectiforms' were sometimes huts, but mostly detailed diagrams of pit-traps and gravity-traps, especially in cases where they were superimposed on a bison or mammoth (but why would only this tiny area depict such devices?). Other motifs were assumed to be lassos, nets and so forth.[10] Occasionally a more exotic ethnographic analogy was used, as where some signs were compared to the traps found in the East Indies that are intended for malignant spirits, for the souls of the dead, or for the spirit of game.[11]

Needless to say, the subjectivity, presupposition and wishful thinking which permeated this theory led to many errors: one of the most blatant concerns the clay bear of Montespan which, from the start, was presented as the epitome of the 'killed' image; the advocates of hunting magic claimed that they could see holes in the bear which meant that it had been

Fig. 99 *'Tectiform' signs painted on a polychrome bison at Font de Gaume (Dordogne). Probably Magdalenian. Bison about 1.6 m long. (JV)*

stabbed with spears or struck with missiles in some sort of ritual attack. Despite the fact that these 'holes' were simply the natural texture of the clay,[12] that the bear is only 1 m from the cave wall, and that the ceiling descends to a very low level immediately beyond it, one still finds dramatic verbal and graphic reconstructions of the scene in popular books, with warriors dancing round a bear head/hide draped over the dummy, and throwing spears at it! These ridiculous flights of fancy are the legacy of the hunting magic theory, whose proponents saw what they wanted to see.

Had they been a little more objective about the data, at Montespan and elsewhere, they would have realised that they were constantly stretching and adapting the theory to fit the evidence, or carefully selecting the facts to fit the theory. For example, faced with the rarity of carnivores in parietal art, they claimed that these were dangerous animals, difficult to hunt, and in any case would not have been considered very tasty. Yet one would think that hunting magic would be most useful for dangerous animals! In any case, carnivores are not so dangerous (they tend to keep well away from humans most of the time), and bulls or stags can be far more aggressive; lions are caught and eaten in Africa, and bears and dogs are considered delicacies in some parts of the world.[13] Besides, how can one reconcile that explanation with the similar rarity of reindeer or birds in parietal art, resources which were of great economic importance at times? Here, the proponents claimed that it was unnecessary to depict these species because they were in plentiful supply and easy to obtain! The inconsistencies in logic are blatant.

Overall, there are very few Palaeolithic animal figures with 'missiles' on or near them: bison seem to have the most, but less than 15% of bison are marked in this way; many caves have no images of this type at all.[14] These marks could be all manner of things (see other theories, below); even if they were missiles, we cannot tell whether they were drawn at the same time as the animal, or years or centuries afterwards: in some cases, such as a pseudo-arrow under the neck of a reindeer from la Colombière, the 'missile' was engraved before the animal![15] One of the most famous examples is a rhinoceros on a pebble from the same site which was thought to be pierced by a number of feathered projectiles: yet such missiles are weak and aerodynamically unstable, and certainly could not penetrate rhino hide.[16]

Similarly, the marks at the mouths and/or nostrils of animal figures are open to many other interpretations, such as voice, dying breath, or breath to show that the animals are very much alive (and see below, p.158) – they have a variety of forms, and are mostly drawn on bison, bears and horses.[17] Supposedly 'falling' animals are often simply drawn on vertical supports, which dictated their orientation.

Besides, what is one to make of 'missiles' associated with human or humanoid figures? The classic example is the engraving of two women from Isturitz, one of whom has a harpoon-like motif on her thigh, which is identical to a mark on the bison engraved on the other side of the same bone (Fig.101)! A little ibex in Niaux appears to have lances sticking out of it, but the same is true of the man at Sous-Grand-Lac, or the humanoids of Cougnac and Pech-Merle (Fig.102).

Even if the figures marked with 'missiles' were symbolic killings, there are many possible reasons for them: they could be mythological, perhaps, or some sort of sacrifice. Sympathetic magic is just one possibility. However, most scholars now dismiss the labelling of these motifs by what they look like to our eyes, and avoid words such as 'arrow' in favour of more objective terms, descriptive of shape (eg chevron with a median longitudinal line).

Breuil and others believed that the carnivores were drawn so that the artist might gain their strength, their skill in catching game, or perhaps be protected from them – a mixture of envy and fear; but in that case, why are carnivore images so rare? Clearly, hypotheses of this kind are untestable. In the same way, opponents of hunting magic argued that many figures were far too detailed and perfect for the job, since presumably a simple outline would have done; but adherents of the theory countered that the better the portrait, the more effective the magic! The arguments go around in circles and lead nowhere.

But if hunting magic is no longer upheld as the primary motivation, Palaeolithic art is still seen as a 'hunter's art' by some scholars who, even today, take some of it at face value as a narrative. This is by no means a new concept: as we have seen, many signs were interpreted as objects associated with the hunt, and the scene of the wounded bison and the man in the Lascaux shaft (Fig.118) has often been considered a depiction of a real event, to the extent that a search was made for the unfortunate hunter's grave nearby!

Proponents of 'hunting art' pick out the very few examples of what they consider to be hunting or stalking scenes, dangerous situations, and animals which are dead, wounded, alarmed or bolting, and present them as if they were typical of Palaeolithic art as a whole;[18] however, not only are these extremely rare, but their interpretation is by no means straightforward, and we have already seen (pp.132-3) that some of the 'groups of hunters' may really be birds, and that the 'dying' or 'wounded' bison of the Altamira ceiling may actually be giving birth or rolling in the dust. The art is considered to have been made by men, and to reflect their obsession with game animals and the hunt, rather like the endless pictures in modern American hunting magazines.[19]

A similar view, which follows on from the idea mentioned earlier (p.124), that the deer figures at Lascaux and elsewhere are accurate records of prize stags, is to see the whole thing as 'trophyism': ie the hunting and display of trophies by men in order to impress and get the girls. The demonstration and advertisement of their hunting prowess, with the prey's attributes emphasised through twisted perspective, improved their status and thus their reproductive success.[20] Palaeolithic art was thus equivalent to a collection of scalps or medals. The idea is appealing, but does not stand up to scrutiny; it does not explain the enigmas in the art such as 'incomplete' or composite images, the large non-figurative component, and the layout in caves.

The suggestion that the organisation of figures within the caves was arranged according to trophy values makes little sense: in that case, one would expect to find ferocious carnivores, bulls and prize stags dominating the principal panels, which is not the case. However, one critic of Leroi-Gourhan's statistical approach to cave-layout (see below, p.168) did confirm that there seem to be three groups of figures: the large herbivores (horse, bison, aurochs), the smaller herbivores (deer, ibex) and the dangerous animals (rhino, bear, feline); rather than interpret them in terms of some metaphysical plan, however, he suggested that their distribution in the caves reflected their economic categorisation by a

Figs. 100/101 *Engraved bone from Isturitz (Pyrénées Atlantiques): on one side there are two females, one of whom has a barbed sign on her leg. The other side depicts a bison with a barbed sign on its side. Magdalenian. Total length: 10 cm. (JV)*

Fig. 102 *Humanoid apparently 'pierced by missiles', in the cave of Pech Merle (Lot). Note the strange 'aviform' sign above. Probably Solutrean. The humanoid is 50 cm high. (JV)*

hunter, based on the amount of meat they provided and the risk they involved.[21]

Another interesting hypothesis, based on a literal reading of the art, is that of corralling: once again, it is by no means a new idea, and the strange circular images drawn at La Pileta, with double finger-marks dotted around inside them, have often been interpreted somewhat tenuously as corrals with animals (or ruminant tracks) inside them[22] (even though identical marks are also found inside animal figures in the cave), while animals with signs painted on or near them, at Marsoulas, Castillo, Niaux and elsewhere, were thought by some authors to be in enclosures.[23]

Recently, however, an expert on animal drives in North America has proposed that much of the painted decoration of Lascaux constitutes a diagram depicting an animal drive, with the rows of dots and the rectangles representing the drive-lines (made of small piles of rocks or similar), blinds and corrals.[24] The barbed signs near some animals are brush barriers which help to funnel the animals into a corral. The little horses are heading for one enclosure, the 'jumping cow' is skidding to a halt in front of another, while the falling horse is going over the edge. This is not the first time that the Lascaux signs have been interpreted in these terms: the quadrilateral images have previously been seen as Texas-style gates for shutting in cattle and horses;[25] but the new hypothesis treats the cave itself as a symbolic drive, forcing the animals down the narrow passages to the 'fall' at the end; the round Hall of the Bulls was perhaps conceptualised as a pound, and the cave itself could be a memory-device to preserve knowledge of how and where to place the cairns and corrals.

Similarly, the round chamber at Altamira is seen as a symbolic pound, with a drive depicted on the ceiling;[26] the curled-up bison at the centre are already dead, while those around them stand and face the hunters (the male humans engraved at the edge of the group). The Altamira ceiling has previously been interpreted as animals falling down a cliff,[27] but never before as a symbolic pound as well.

The hypothesis has the merit that it concerns a technique which is certainly the most efficient way to exploit large herbivores, and which was surely used throughout the Upper and probably the Middle Palaeolithic; it is also solidly based on ethnographic information, including the methods of depicting such drives in other cultures. Indeed, the interpretation seemed self-evident to the expert on animal drives as soon as he saw these caves, so close were the analogies to ethnographic examples. It is not claimed that all the 'signs' are corrals or drive-lines, but their close association with the animals at Lascaux certainly make his hypothesis thought-provoking.

In the last two decades we have become so accustomed to esoteric interpretations of Palaeolithic art that we tend to forget that not all of it is necessarily mysterious and metaphysical: much of it may well be mundane and deal with everyday concerns. Certainly some aspects of animal exploitation seem to be present: for example, there are one or two depictions of what may be stripped carcasses, or of animal hides spread out;[28] and the earliest art on blocks includes a series of what appear to be animal prints (probably a bear's front paw – a large cupmark with an arc of 3–5 little ones above);[29] other motifs which have traditionally been interpreted as vulvas may also be horse-hoof or bird-foot prints,[30] while the series of double incisions on the handle of the carved Lascaux lamp could

Fig. 103 *Carved antler (possibly a spear-thrower) from Le Mas d'Azil (Ariège) showing three horseheads, one of them apparently skinned. Magdalenian. Length: 16.3 cm. (JV)*

be the prints of an ungulate such as a boar. In any case, it is certain that the observation and recognition of animal and bird tracks were of great importance in Palaeolithic life, and it is no surprise that they were occasionally reproduced.

However, even if hunting played a role in the production of some Palaeolithic art, for any or all of the reasons outlined above, the fact remains that the art's content is far more complex. We have already seen (Chapter 4) that the species depicted cannot be taken as an accurate record of what was present in the environment; and it is equally true that they often bear little relation to the use people made of what was available: ie to the faunal remains in occupation sites.

It is often the case that we do not obtain an adequate picture of diet from the bones, because of differential preservation, or because the sample is small; and, as we have seen, it is often theoretically possible that the people who ate the animals were not from the same period as the artists, though in most cases they probably were.

However, the results from many sites suggest that depicted species rarely correspond to eaten species in terms of quantity. This applies not only to parietal but also to portable art, and was noticed by some of the earliest specialists such as Cartailhac. For example, at Gönnersdorf the mammoth is the second most frequently depicted animal (after the horse), but is very rare in the fauna, while the reindeer, which abounds in the fauna, is absent from the art.[31] At Dolní Věstonice, the mammoth dominates the bones but is rare in the art, which is primarily devoted to carnivores.[32] At La Vache, the ibex is overwhelmingly predominant in the fauna but only accounts for about 18% of the depictions, while the horse, which predominates in the art (over 25%) represents only 0.01% of the fauna![33] The bones from the grotte des Eyzies and the abri Morin are dominated by reindeer, but their portable art is primarily devoted to horses.[34] On the other hand, both the slabs and the bones of Limeuil are dominated by reindeer, and the horse dominates in both at Gönnersdorf.

In parietal art, the example always given is Lascaux, where reindeer account for 90% of the bones but were drawn only once (0.16%); as mentioned earlier (p.124), it is possible that many of the cave's stag figures may actually be reindeer, which would alter this disparity a little; but horses still account for 60% of the figures. Bison, aurochs and ibex are in the cave's art but not in its bones; this situation is reversed for boar and roe deer.

The same picture is repeated elsewhere: the ibex occurs in the art of Pair-non-Pair but not in its fauna; at Comarque, over 75% of the bones are reindeer, but the figures are mainly horses;[35] at Villars the people ate reindeer, but did not draw any; in the Ardèche caves, the mammoth predominates in the art (almost half of the figures) and the ibex and reindeer are almost absent, whereas the bones show an abundance of the last two species and an absence of mammoth; at Gargas there are no reindeer figures although over a third of the bones are of that species, whereas the mammoth, absent from the fauna, was depicted six times.

The situation recurs in Cantabria: at Altamira they drew bison and ate red deer; at Ekain the 34 horses are 57.6% of the animal figures, but there are almost no horse-bones in the fauna which is dominated by ibex and red deer; while at Tito Bustillo, the fauna is heavily dominated by red deer (up to 94%) with ibex second and very few horse bones, whereas the parietal art has 27 horses (37.5%), 23 red deer (31.9%) and 9 ibex (12.5%).[36] On the other hand, the red deer dominates both the art and the bones at Covalanas and at La Pasiega.

In short, the relationship between depicted and eaten species ranges from non-existent to quite good; once again, there is no rule, no homogeneity. Nevertheless, it remains true that the restricted range of species depicted has no direct equivalent in any excavated level's sample of bones, and it is clear that the motivations behind the art were different from the environmental factors and economic choices which produced the faunal remains. A similar situation occurs in Gallo-Roman sites, where the animals depicted on pottery are over 75% wild, with dogs dominating the other 25%, whereas around 90% of the bones come from domestic species.[37]

The disparities have been presented in the form of cumulative graphs which demonstrate, for example, that art and bones vary far more widely at Lascaux than at Gabillou.[38] Another recent study, however, which compared the combined percentages of depicted species in 90 French caves with those of bones from 151 French sites spanning the entire Upper Palaeolithic, and those of 35 Spanish caves with the bones from 30 Spanish sites, was a waste of effort:[39] it only makes sense to compare an individual cave's figures with the bones from the same site, or at least from the same locality and period, since climate and local topography played a major role in economic choice – for example, ibex dominate the fauna in rugged ibex country, the other herbivores in flatter areas.

Although we can accept that the practice of sympathetic magic and the actual depiction of hunting may account for a small amount of Palaeolithic art, the evidence for them is extremely scarce. However, there is a related phenomenon which could lie behind rather more of the art: shamanism. Many early scholars, as we have seen, interpreted the 'composite' figures as sorcerers or shamans in masks, basing their argument simply on similarities to ethnographic depictions and accounts. Today, the theory survives in connection with particular hypotheses: for example, if the Lascaux animals are indeed being driven, then the enigmatic 'unicorn' figure, sometimes interpreted as men in a skin, may be a shaman driving the little horses before him.

The shaman is a very important figure, being a person with spiritual powers who combines the roles of healer, priest, magician and artist, as well as poet, actor and even psychotherapist! His most important function is to act as liaison between this world and the spirit world, a task usually performed by means of trances and hallucinatory experiences – either self-induced or using hallucinogenic fungi or other similar substances. Cultures with shamans usually have a zoomorphic view of the world, and things are seen and experienced in animal form. Hence, from this perspective, Palaeolithic images are 'spirit animals', not copies of the real thing, while the shaft-scene at Lascaux becomes a shamanistic fight, a psychic conflict between two shamans (one in animal form) or between a shaman and a malevolent spirit.[40]

In a theory of this kind, based on Siberian ethnography, Glory suggested that many of the figures in Palaeolithic art were 'ongones', spirits which took the form of 'zoomorphs', 'anthropomorphs' and 'polymorphs', and which were asked to help in hunting, matters of health,

Fig. 104   *Macro-photo of the head of a composite engraved figure in Trois Frères (Ariège), together with Breuil's tracing of the full figure. Probably Magdalenian. Total length: 30 cm. The head is about 3 cm across. (JV, collection Bégouën)*

and so on. Here, the lines emanating from animals' mouths would be an evil 'illness spirit' being exorcised, while damaged images will have been broken in anger when prayers were not answered.[41]

A closely related view, based on concepts widely held among modern hunting peoples, concerns the 'master-of-animals', usually a dead shaman; this humanoid or composite figure represents a third force mediating between living shamans and animals, one which constitutes the life-force of an animal or which can impart life-force to it. The living shamans derive their own life-force from the animals and then use it in the service of their clients.[42] According to this scenario, the artists were shamans maintaining links with the animals (with which they closely identified) through the master-of-animals; the power derived from the art came through the act of drawing, not from subsequent viewing. The lines at the mouths and nostrils of animal depictions represented the entrance or exit of life-forces.

A different possibility is that these lines represent the nose-bleeds often experienced by shamans entering a trance – this is certainly a less fanciful explanation of the marks in front of one Trois Frères composite than the traditional one of playing a 'musical bow'! It is worth noting that what appear to be lances or missiles hitting animals' bodies may likewise have an alternative explanation: eg an Apache myth concerning the creation of the horse involves four whirlwinds penetrating the animal's flank, shoulders and hips, enabling it to breathe and move.[43]

Obviously, the shamanistic explanations, like the similar ones based on totemism in other cultures, involve ethnographic parallels; but from what we know about Upper Palaeolithic culture and life (particularly the Magdalenian), it seems likely that beliefs of this kind existed and thus played a role in the production of Palaeolithic art. Certainly one of the major directions of current research, involving hallucinations, trances and 'phosphenes' (see below, p.190), is derived largely from the study of the shaman phenomenon in southern Africa and its role in the production and content of San rock-art.[44]

Fertility magic and sex

One of the most popular and durable explanations for much Ice Age art has been that it involves 'fertility magic': ie that the artists depicted animals in the hope that they would reproduce and flourish to provide food in the future. As mentioned earlier, such utilitarian theories about Palaeolithic art were largely inspired by an early study of the Arunta of Australia who, it was said, drew animals in order to make them multiply.

The proponents of this theory, like those of hunting magic (and they were often the same people), selected examples which seemed to fit their ideas, and they often saw what they wanted to see: animals mating, and an emphasis on human sexuality too. Yet, as already mentioned, relatively few animals have their gender shown, and any genitalia depicted are almost always shown discreetly. There are no pictures of rampant stallions, and the only known animal with a blatant erection is a bison at Trois Frères which is probably a composite, and thus probably at least half human (though the phallus is thought to be that of a bison).[45]

As for copulation, in the whole of Palaeolithic iconography there are only a couple of possible examples, and they are dubious in the extreme. The best known is an alleged coupling of horses in the sculptured frieze at La Chaire à Calvin (Charente): there may be two horse figures here, partly superimposed, but the 'stallion' is not in the correct position for copulation, and is smaller than the 'mare' (Fig.105).[46] Some scholars believe that the small horse is actually superimposed on a bison figure. The other example, two bison at Altamira, is equally odd, as they are claimed to be a female mounting a (larger) male in order to arouse him sexually![47] The proponent of this view also sees the Altamira ceiling as the depiction of a bison herd in the rutting season.[48]

Fig. 105 *Sculptured frieze of La Chaire à Calvin (Charente) with sketch showing the two superimposed animals. Probably Solutrean. Total length: c 3 m. The superimposed figures are 1.4 m long. (JV)*

Faced with an absence of copulation scenes, proponents of fertility magic have therefore interpreted many pairs of animal figures as being engaged in 'pre-copulatory activity': ie males advancing on females, or sniffing them.[49] Here, too, a great deal of wishful thinking is involved: for example, at Teyjat we definitely have a male aurochs behind a female (Fig.85), and his muzzle is close to her rump – but they are part of a line of aurochs, and there is another cow behind the bull. Similarly, the two clay bison in the Tuc d'Audoubert have often been interpreted as a male about to mount a female: but neither has its sex shown, so that one can only go by relative size; they are not close together, and the 'male' is actually behind and 15 cm beyond the 'female' (Fig.120); and, of course, the interpretation ignores the other two bison here, a small clay model and an engraving in the floor.

The last resort of those seeking evidence of an interest in animal fertility was to see many figures, especially horses, as 'pregnant'. Yet even

Fig. 106 *Two superimposed humanoids on plaquette from Enlène (Ariège), with a third at the other end of the stone. Note also the legs and belly of a bison. Magdalenian. Total length: 21 cm (Photo J. Clottes, collection Bégouën)*

veterinarians cannot tell if a horse is pregnant from its profile alone (though they can do so from a back-view);[50] quite apart from possible stylistic conventions involved in giving horses big round bellies (like short legs, tiny heads, etc), inflated stomachs can also be the result of eating quantities of wet grass; and some male horse figures with phallus shown look equally pregnant![51]

In a couple of cases, a small figure is superimposed on a large animal's body – most notably at Lascaux where a foal appears to be 'inside' a mare – but they are unlikely to be pictures of pregnancy. The Lascaux case has the foal in the wrong position, and one of its limbs protrudes below the bigger figure's belly, so they are far more likely to be two superimposed images or, according to another suggestion, a mare cantering alongside her foal.[52] Besides, as we have seen, there are very few young animals in Palaeolithic art.

So much for animal fertility, therefore. Surprisingly, more attention has been given to human sexuality, as if the artists were using fertility magic to produce their own children. Here again, however, much is in the eye of the beholder.

We have already seen that quite a few human figures can be sexed, though relatively few have their genitalia marked: some are clearly female, fewer are clearly male, and most are 'neutral' – ie asexual. Since some of the definite males have erections, one might expect that intercourse would have been drawn from time to time.

There have been one or two claims for depictions of human copulation, but they are very sketchy, and it requires a great deal of goodwill and wishful thinking to see them as any such thing. For example, some of the 'fat ladies' of La Marche may actually be thinner ladies with men on top of them, and the 'Femme au Renne' of Laugerie-Basse also has faint lines engraved above her which may represent a person.[53] A large plaquette from Enlène, with a bison as its principal figure, also has a couple of superimposed human/humanoid figures which have been interpreted as a male taking a female from behind[54] – yet, apart from possible whiskers on the male, the only indication of gender is a difference in size (the smaller figure's belt and long hair are insufficient evidence), and there is no explicit trace whatsoever of sexual activity (Fig.106).

The only plausible scene of human copulation in Palaeolithic art is to be found in the cave of Los Casares where, according to a tracing by Cabré, a

Fig. 107 *Supposed copulation scene, of dubious date, from Los Casares (Guadalajara). (After Cabré)*

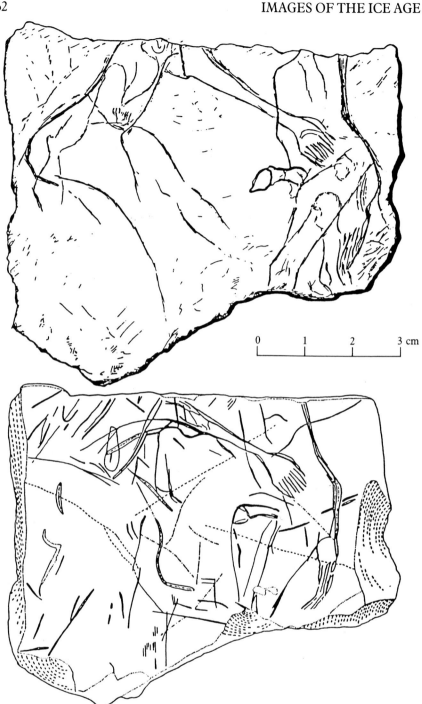

0   1   2   3 cm

Fig. 108 *Breuil's version and a
more recent tracing of a plaquette
from Enlène (Ariège) showing the
differences in content. Magdalenian.
(After Bégouën et al)*

male human has a huge phallus which is superimposed on the stomach of
another human with prominent hips and abdomen, who is probably
female.[55] Even here, however, the scene is not a realistic depiction of
intercourse, and in any case the date of the figures and the accuracy of the
copy are in some doubt (Fig.107).

As with animal figures, the proponents of fertility magic therefore
looked for scenes of 'pre-copulatory' activity: one example, on another
engraved stone from Enlène, has a male human with an erection, and
facing him is a possible second figure with arm outstretched. In Breuil's
original tracing, the second figure was 'completed' into a woman, with
breasts and prominent buttocks, who was thus reaching to grasp the
phallus in her hand; many of these features, however, are simply not
there.[56] The 'woman' is extremely sketchy, and may not even be a figure,

while the 'arm/hand' could equally well be an animal's tail with a tuft of hair at the end; and in any case it 'touches' the man's stomach, not his sex. As the engraving is broken, the man's head is missing, but he may be a composite figure, since he appears to have a furry tail hanging at the back (Fig.108).

Similar subjectivity influenced interpretation of the engraved rib from Isturitz depicting two humans. They were seen as a male pursuing a female, the latter identified by her breasts and the former by a 'gesture' or a glint in the eye! Yet closer inspection reveals that the second figure, broken off at the chest, clearly had breasts too; moreover, the position of the breasts indicates that these are not two figures crawling along a horizontal surface, with one pursuing the other, but two vertical figures, one placed below the other (Fig.100).[57]

What about pregnancies, then? As with horses, any fat female figure was interpreted as being pregnant; but a re-evaluation of 'Venus' figurines (see above, p.138) has suggested that very few of them (17%) could plausibly be interpreted in that way.[58] None is definitely shown in the act of childbirth, although one stone figurine from Kostenki, different from all the others, appears to be a pregnant woman about to give birth in the Asiatic position (she is kneeling with legs apart, and her genitals are clearly marked); and gynaecologists have interpreted the Tursac figurine as a woman giving birth, with its 'peg' as the emerging child, and also one Grimaldi figurine and the bizarre Monpazier statuette as being in the act of giving birth.[59] Apart from these rare possibilities, 'Venus' figurines do not, therefore, seem to glorify maternity or fertility; they appear instead to be about womanhood. There are no known pictures of children, apart from the possible baby at Gönnersdorf (see above, p.141), and a few stencils of infants' hands.

Fig. 109   *The 'playing card' figure from Laussel (Dordogne). Probably Gravettian. Size: 20 cm. (JV)*

Finally, an enigmatic carving from Laussel which resembles a playing card, with two half-figures joined at the waist, has often been claimed to be either a copulation scene or a woman giving birth; but it is so vague that it could be anything (Fig.109):[60] indeed, some scholars see it as a full Venus figure, while another view argues that it is a person standing waist-deep in water!

The obsession with sexual interpretations, and particularly with female genitalia, in the art can be traced back 75 years. In 1911 the abbé Breuil was consulted about certain deeply engraved motifs which had been found on some stone blocks in early Upper Palaeolithic sites in Dordogne. Breuil declared these ovoid and subtriangular figures to be '*pudendum muliebre*', or vulvas.[61] This was a highly subjective interpretation but, ever since then, most scholars have accepted it without question – even though a man of Breuil's profession should not perhaps be considered an expert on this particular motif! – and it was certainly a major factor in Leroi-Gourhan's concept of the development and complementarity of male and female signs (see below, p.169).

The 'identification' of so many examples of female genitalia led to ideas about a Palaeolithic obsession with sex; there is tautology in the reasoning – the figures are assumed to be vulvas, from which an obsession with sex is inferred, the evidence for which is the vulvas! Independent data are required to support the hypothesis; but when one searches, vulvas are remarkably hard to find in Palaeolithic art. If one allows that the only *definite* vulvas are those found in context, that is, in full female figures, then they are very few in number; of the motifs without context, only the vulva modelled in clay at Bédeilhac (see chapter 5) can be identified with any confidence – all the rest are mere interpretations.

It is surprising to find that, even among the 'Venus' figurines, often seen as proof of an intense interest in female sexuality, very few have the pubic triangle marked, and even fewer have the median cleft.[62] In fact, these figures accentuate the breasts, buttocks and hips; they may represent fecundity but they do *not* draw attention to the vulva, unlike the headless bas-reliefs of Angles-sur-l'Anglin which do seem to have this part of the body as their focus.

Where vulvas are depicted, nearly all are triangles, as one finds in female figures from many other periods and cultures. Yet the early Upper Palaeolithic motifs have a wide variety of shapes, and scholars have needed a number of excuses in their desperate bid to fit all these shapes to the chosen interpretation: eg there are 'incomplete vulvas', 'squared vulvas', 'broken, double vulvas', 'circular vulvas', 'relief vulvas', and even 'trousers vulvas'! Others have claimed unfinished vulvas, atypical vulvas, and even interpreted a single straight engraved line as an isolated vulvar cleft! (Fig.54)[63]

But in fact, as we know from rock-art in Australia and elsewhere, very simple graphic designs can have a very wide range of meanings; and in any case, if the Palaeolithic artists took pains to differentiate these shapes, it makes little sense for us to lump them all together into a single, subjectively chosen interpretation. Hence, more objective scholars have divided them into descriptive rather than interpretative categories: horned ovals, pear-shaped ovals, arched ovals, etc.[64]

Wishful thinking, therefore, permeates the search for human sexuality in Palaeolithic art. For example, the clay 'sausages' in the Tuc d'Audoubert were automatically seen as phalluses by believers in fertility magic, while others saw them as possible bison horns, but, as mentioned earlier (pp.94-5), they almost certainly represent the sculptor trying out the clay.[65] Similarly, one of the portable statuettes of Angles-sur-l'Anglin is a kneeling female, but Breuil originally thought it was a phallus.[66]

It is possible, of course, that there is purposeful ambiguity here, a visual pun; the same may apply to 'vulvas' which look more like horse-hoof or bird-foot prints. We have seen that all boundaries and categories are blurred in Palaeolithic art, and it would not be surprising if the artists noticed and played with such ambiguities.[67] This kind of punning may also explain why they drew a red human outline around each of two little stalagmites, a few metres apart, sticking out of the wall at Le Portel, thus turning them into ithyphallic men (Fig.52).

A different example of ambiguity probably owes more to the preoccupations of the beholder than to the intentions of the artist: a series of little ivory carvings from Dolní Věstonice have always been seen by most male scholars as abstract female figurines with long necks and huge buttocks or breasts; the same applies to an ivory rod from the site, which has a pair of 'breasts' protruding from it, some way up. However, feminist scholars suggest that these carvings should be inverted, and clearly represent male genitalia![68]

Just as the 'hunting magic' theory has developed into a view of the art as reflecting a male preoccupation with hunting, so 'fertility magic' has given way to a 'cheesecake' view of human depictions: ie this is early erotica made by men for their own pleasure. This is why the female figures have exaggerated breasts and buttocks, with no emphasis on the limbs or the face. They represent women's contours and were intended to arouse, to be touched, carried and fondled.[69] Similarly, the act of engraving the alleged 'vulva' carvings of the early phases reflected the 'digital activity of lovemaking around the labia'![70]

It is even suggested that the statuettes' heads are bowed in subservience, submission or shyness, while the enlarged rumps on the schematised females of Gönnersdorf and elsewhere are simply sexual lures. The rarity of clothing and personal adornments in the depictions, as well as of birds and other small resources, proves that it is an art which is limited to male preoccupations, and not concerned with a woman's side of life.[71] In this it resembles modern Eskimo art, comprising hunted animals and a few nude females.

Finally, some Palaeolithic depictions of women have been compared with pin-ups in *Playboy*, and it has been claimed that 'female figures often [*sic*] appear in sexually inviting attitudes, which may be quite the same as those in the most brazen pornographic magazines. There are also anatomically detailed pictures of the vulva, showing the female sex organ sometimes frontally, sometimes inverted and from the back, open to penetration'![72]

This quotation reveals a lurid imagination. Apart from the bas-relief reclining females of La Magdeleine, there are no such 'poses' in Palaeolithic art. A few depictions do bear a superficial resemblance to the pin-ups selected, but the range of positions in which the female body can be depicted is fairly limited, as shown by the fact that all the comparative material was found in a single issue of the magazine! Another basic point is that we do not know the sex of the artists. The carvers of the 'vulvas' and figurines could just as easily have been female, and one can extend this argument to the whole of Palaeolithic art, invoking initiation ceremonies to explain menstruation, with lunar notation (see below, p.182) as supporting evidence.[73]

The theory that the art is about the male preoccupations of hunting, fighting and girls comes from twentieth-century male scholars. Feminist scholars have a different point of view about the supposedly sexual images (for the same reason, it would be interesting to see the reaction of a vegetarian culture to the animal figures). The macho view of Palaeolithic art is both simplistic and anachronistic. Some of the female figurines may have aroused men, but there is no reason to suppose that this was their intended function: after all, many of them were carefully hidden in pits. Moreover, images of female genitalia are far rarer than has been claimed in the past, and those which were depicted may well have been intended simply to indicate gender, rather than be erotic. The greater part of Palaeolithic art is clearly not about sex, at least in an explicit sense. The next major theoretical advance, however, introduced the notion of a symbolic sexual element.

Laming-Emperaire and Leroi-Gourhan: the female and the male

Despite his stubborn opposition to parietal art, Emile Cartailhac was a very perceptive man; we have already mentioned (see above, p.150) his early impression that 'art for art's sake' was an insufficient explanation. A further example occurred in 1902, when he stated that the cave of Pair-non-Pair appeared to contain a random accumulation of pictures, but that of La Mouthe had been decorated systematically.[74]

Thanks to the dominance of Breuil, who treated cave-art as a collection of individual pictures, this pioneering insight was forgotten, and the concept did not re-emerge for over half a century, when two scholars, influenced by the unfinished work of Max Raphael, arrived more or less independently at the same conclusion. Annette Laming-Emperaire and André Leroi-Gourhan, in a series of publications between 1957 and 1965, revolutionised the study of Palaeolithic art, which thus took its first leap forward since 1902.

One of the basic features of their approach was to ask new questions, while avoiding the intuition and the subjective use of ethnographic parallels which had characterised previous work and led to an impasse. Interpretation was to be based on the total corpus of figures in a cave, not on one or two selected images. They treated parietal art as a carefully laid-out composition within each cave; the animals were not portraits but symbols; they did not reflect explicit practices but rather a complex metaphysical system or 'mythogram'. Previous concentration on the aesthetic and magical aspects was inadequate, as the ideas behind Palaeolithic art were far more complex and sophisticated. In short, they moved from analogy to observation, from naive and particularist explanations to the structures reflected in the subject-matter and its associations. Although there was no hope of grasping the art's meaning, one might gain some insight into the richness of thought behind this symbol system.[75]

Fig. 110 *The 'central composition' of black figures at Santimamiñe (Vizcaya), showing a horse surrounded by a number of bison, including two drawn vertically, back to back. Probably Magdalenian. Panel size: 2.5 m. The horse is 45 cm long. (JV)*

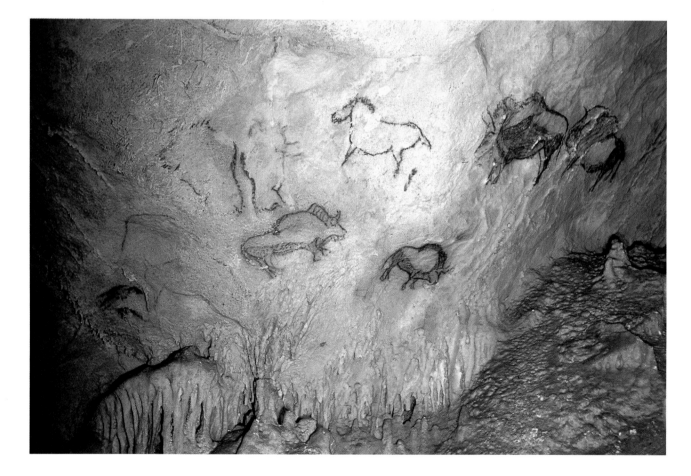

Although it was Laming-Emperaire who, following Max Raphael, first treated parietal figures as a composition in her study of the sculptured sites, as well as Lascaux and other caves, it was Leroi-Gourhan who confirmed the notion, extended it to all major caves and continued to develop the idea, while Laming-Emperaire eventually dropped it in favour of an even more complex explanation (see below).

An investigation of how many figures of each species existed in each cave, together with their associations and their location on the walls, led to a number of insights: Leroi-Gourhan eventually divided animal figures into four groups,[76] based on the frequency of their depiction in parietal art as a whole. Group A was the horse, which constitutes about 30% of all parietal animals; B was the bison and aurochs, which also account for about 30%; C was animals such as deer, ibex and mammoth, representing another 30%; D, the final 10%, comprised the rarer animals such as bears, felines and rhinos.

He also divided caves into entrance zones, central zones, and side chambers and dark ends. It appeared that about 90% of A and B were concentrated on the main panels in the central areas, the majority of C figures were near the entrance and on the peripheries of the central compositions, while D animals clustered in the more remote zones. He developed the concept of an ideal or 'standard' layout to which the topography of each cave was adapted as far as possible (Fig.111): these were organised sanctuaries, with repeated compositions separated by zones of transition marked with the appropriate animals or signs.[77] Different signs are found at the start and end of the sanctuary (mostly dots and lines), while more elaborate signs occur before and within the big animal compositions.

Fig.111 *Leroi-Gourhan's 'blueprint' for the ideal cave layout. (After Leroi-Gourhan)*

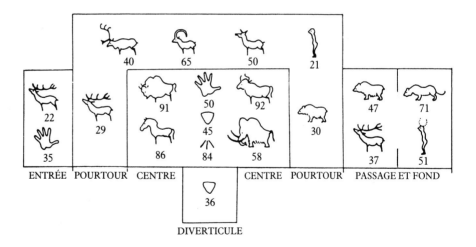

However, as Leroi-Gourhan was well aware, many caves are extremely complex topographically and may have had different entrances during the Palaeolithic which are now blocked: one cannot simply assume that the present entry has always been the only one. Our doors may have been their windows! From the start, he sometimes had to adopt bizarre itineraries within caves in order for his scheme to fit – even going to the wall furthest from the entrance to find the 'first series' of big signs, and then coming back again![78] He knew that there is no 'average' cave but still believed that, though topography may have dictated layout to some extent, it was still the 'mythogram' that conditioned the choice of topographic adaptation;[79] ie he imagined that the artist who had been 'commissioned' to decorate the cave checked its topography and reliefs, and noted any features which appeared useful in accordance with the predetermined blueprint – hence the Altamira ceiling-bosses could become only curled-up bison, not felines or something else.[80]

His divisions into central and peripheral areas are somewhat tenuous (those in Gabillou, for example, are quite arbitrary – where do complex panels start and end?) and decorated caves are far more heterogeneous than the scheme suggests. It is not just a question of shallow caves and rock-shelters versus simple corridors and complex systems of galleries, but a problem of content: as we have seen, some caves have a single figure or sign, others have animals and no signs or vice versa;[81] and how is one to fit the newly discovered open-air engravings into this concept? Leroi-Gourhan himself admitted that he omitted caves that were exceptions, incomprehensible or lacking the appropriate animals; and he conceded that there were different regional 'formulas', notably in Cantabria.[82] Far from presenting his theories as impeccable, he constantly pointed out the gaps and faults in them.

Even in caves with abundant animals and signs, the latter do not always begin or end the decoration, and the former often do not conform to the scheme: for example, bison are the last animals in the 'feline gallery' at the end of Lascaux, and among the last in Les Combarelles, while the clay bison of the Tuc d'Audoubert are at the far end of the cave. Leroi-Gourhan sometimes assumed changes in cave topography in order to explain 'anomalies' of this kind (eg wrongly claiming that the Tuc d'Audoubert was once joined to Trois Frères, so that the clay bison would not be at the end) or divided the caves up on the basis of the themes represented rather than topography, thus turning his scheme into a circular argument.[83] Even so, his own figures revealed that 60% of bears and 66% of rhinos are in the 'central' areas; indeed, the rhino is rarer in cave-ends than are ibex and deer! Ibex can be found in virtually any part of the cave. It is true, however, that the D group does not usually occur in entrance zones, and that felines tend to be in remote zones (a fact already noticed by Breuil).

A further problem with Leroi-Gourhan's scheme is his use of quantitative data and percentages (which often contain small errors[84]), and especially the fact that the scheme works on a presence/absence basis, not on abundance: hence a single horse figure is the equivalent of a mass of bison, and vice versa. Few scholars have been able to accept that abundance was irrelevant, quite apart from other variations such as colour, size, orientation, technique and completeness. Leroi-Gourhan pointed out that, in Christian churches, images of Christ are often far outnumbered by the Madonna and saints, but admitted that the 17 hinds of Covalanas probably had a more important role than the single (and dubious) aurochs head in the cave.[85] A number of caves are similarly dominated by a single species.

Yet in his final presentation of the scheme he allows A/B/C/D animals to be present in any proportions in a particular cave. Where the C group is concerned, only one of the species need be present in the formula. Caves with no A animals (Tête-du-Lion) or with no A or B (Cougnac) are seen as exceptions that prove the rule![86]

In fact, there are so many exceptions, contradictions and variations that one can seriously doubt the validity of the 'ideal layout'. Recent detailed studies, both of individual caves and of regional groups, constantly stress that each site is unique and has its own 'symbolic construction' adapted to its topography and concretions, and with multiple thematic linkages.[87] Few people would now uphold the universality of the layout, either in space or time. Leroi-Gourhan lumped all Palaeolithic parietal art into his scheme, and believed that its structure remained homogeneous and continuous for 20,000 years – even in caves decorated in several phases, the later artists respected the first sanctuary's layout.[88] There is certainly a degree of thematic uniformity over this timespan: caves are decorated with the same fairly restricted range of animals in profile, and seem to represent variations on a theme: as Leroi-Gourhan pointed out, nobody drew a frieze of lions and storks surrounded by hyenas and eagles![89] He therefore thought that the animals depicted corresponded to a certain mythology.

The other foundation stone of the approach was the discovery of repeated 'associations' in the art, and the claim that there was a basic dualism. This, too, was a concept first developed by Laming-Emperaire and subsequently extended by Leroi-Gourhan. Having found bison 'associated' with both horses and women, Laming assumed that the horse might in some way be equivalent to female, and thus the bovids with the male. Leroi-Gourhan, on the other hand, saw the bison as equivalent to females; his view was based almost entirely on a couple of crude figures in Pech-Merle which he interpreted as human females derived from bison with raised tails (Fig.112).[90] However, the animal figures are vague (they could even be horses!) and any similarity is superficial, being limited to the curve of the profile: in short, the link between females and bison is so tenuous as to be non-existent.[91]

Laming also found a frequent association of horses with either bison or aurochs, a fact which Capitan and Bouyssonie had already noted in the portable art of Limeuil in 1924, though in caves it is of little surprise in view of their great abundance. In the new approach, the numerically dominant horses and bovids, concentrated in the central panels, were thought to represent a basic duality. One possiblity was that this was sexual, though it was admitted that other explanations might be equally plausible (eg sun/moon, heaven/earth, good/evil, or something more complex). It was pointed out that horses live under the control of a stallion, and cattle under that of cows.[92]

Having tentatively claimed a sexual duality, it was natural for Leroi-Gourhan to extend this to the abstract signs; his somewhat Freudian viewpoint was that they could be divided into male (phallic) and female (vulvar) groups, which he later dubbed 'thin' and 'full' – a third group was later added to include the dots. This view was inspired by two bear-heads in portable art, one of which (La Madeleine) seemed to be associated with a phallus, and the other (Massat) with a 'penniform' sign.[93] Hence, where proponents of hunting magic saw missiles and wounds, Leroi-Gourhan saw phallic and vulvar symbols.

In the early years of his research, he felt certain that the two series of signs were of a sexual nature; but it was never explained why such a wide range of shapes should have only two basic meanings, even allowing for artistic licence. Besides, there were inconsistencies such as the claviform which is thin but, as we have seen (p.142), was thought to be derived from a female silhouette – in the same way as the Coca Cola bottle could be taken either as a phallic symbol or as a stylised female form, according to one's preference!

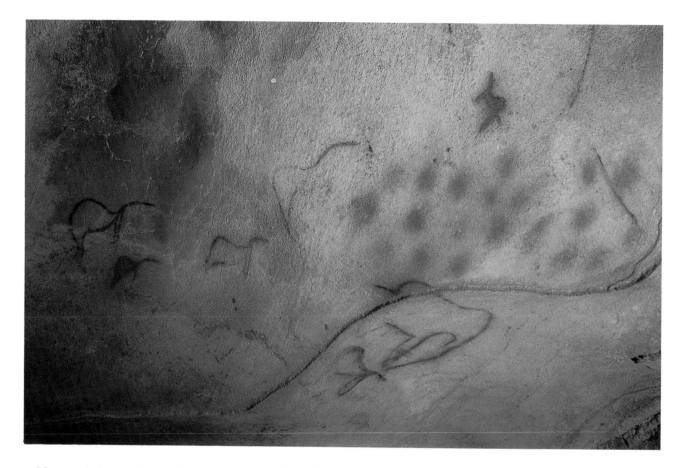

Fig. 112   The 'bison-women' of
Pech Merle (Lot). Probably
Solutrean. Each about 10 cm long.
(JV)

Fig. 113   The art in its setting: dots
and a hand stencil in Pech Merle
(Lot). Probably Solutrean. (JV)

Many scholars rejected the sexual theory: Breuil, for example, referred to it as a '*tentative déplorable*' and a '*perspective sexomaniaque*',[94] which was somewhat ironic coming from the man who had unleashed the obsession with vulvas in 1911! However, most critics were confusing sexuality with genitalia; even if the division existed, it probably had nothing to do with eroticism, but simply reflected some sort of structure in society.

From these horse/bison and thin/full oppositions and associations, Leroi-Gourhan developed a view of the caves in which other animals were allotted sexual roles (eg ibex were male, mammoths female). Even clearly male bison were considered female symbols – in other words, animals were sexed by species and location, not by gender (except for stags and hinds, for some reason)! By 1982, however, he had dropped the emphasis on sexual symbolism, although he retained the entirely valid point that the cave itself probably constituted a female (womb) symbol, as it has in many subsequent cultures.[95]

Just as his scheme had fixed places for different species, with 'female' animals and signs clustering in central areas, and male elements both among them (horse) and around them, it also claimed that human males were mostly at the back of the caves, while females were in the well-lit entrance areas (as at La Magdeleine, etc). Occasionally he allowed the theory to dictate the evidence, to the extent that humanoids were identified and gender attributed to unsexed humans/humanoids, on the basis of their location in the caves! For example, the '*Femme à l'anorak*' of Gabillou was claimed to be a woman because of its position, but he ignored the fact that the cave's clearest female is right at the very end.[96]

His claim that human males are always at cave-ends or on the periphery of panels was untenable because he was including many unsexed heads, and he himself admitted that males can be found all over the place – in entrance zones, cave-ends or around central panels.[97] In fact, there is a clear patterning of human/humanoid representations in caves, but it is somewhat different from that proposed by Leroi-Gourhan. Applying strict sexing criteria (see above, p.136), and dividing caves into daylight, obscurity and total darkness, it has been found that the percentages of definite males, definite females and neutrals change dramatically: in daylight zones, 56.2% of humans are female, 6.2% male and 37.5% neutral; in obscurity these figures have changed to 31.4%, 8.1% and 60.4%, while in total darkness only 3.6% of the humans are female, 16.3% are male and a massive 80% are neutral.[98]

Leroi-Gourhan felt that his formula comprised a variety of binary oppositions, and that a symbol such as the horse or bison could certainly have taken on or shed many associations in the course of the Upper Palaeolithic – he never saw the art as a rigid, fossilised system. Some of the associations which he and Laming discovered are certainly real, although the terms 'juxtaposition' or 'coexistence' might be preferable since they do not assume that the phenomenon is meaningful: for example, many hand stencils are found in close proximity to painted dots (eg Pech Merle (Fig.113), Castillo), and often near horses too; over half of parietal signs are with animal figures; in parietal art as a whole, bison and aurochs are found with horses far more often than with any other species. In portable art, however, the same rules do not apply: at La Marche, horses are found with practically every species depicted, while humans are found with almost every species except felines.[99]

The overwhelming parietal coexistence of horses and bison suggests that their duality is the basis of the whole figurative system, as Leroi-Gourhan believed; alternatively, these animals may each have a separate importance, making their frequent coexistence a matter of chance (the limited range of species depicted also increases the chances of their being found together). We simply do not know. However, a recent re-evaluation of parietal figures and their associations, using factor analysis, not only confirms the fundamental role and opposition of horses and bison, but also suggests that the bison is dominant, since it has a greater 'effect' on a panel centred on a horse than the presence of a horse does on a bison panel.[100]

It was also found that some 'associations' were rare or non-existent – such as bison and aurochs, bison and stag, or aurochs and reindeer; however, some species are rarely associated because of regional differences: as mentioned earlier (p.133) the majority of hinds occur in the early Magdalenian of Cantabria, while mammoths are predominantly found in the mid-Magdalenian of Dordogne – this considerably lowers their chances of being found together.[101]

Other scholars have run statistical tests on Leroi-Gourhan's figures but found little evidence to support the duality of signs or the male/female dichotomy in animal species, especially the homogeneity of male signs and animals, though the 'female animal' and 'dangerous animal' groups remained plausible.[102] Others claimed that Leroi-Gourhan's use of percentages of associations was 'idiosyncratic and fundamentally unsound' because he ignored group size and relied simply on presence/ absence. A multivariate analysis of the animal figures in a number of sites revealed two small groups of neighbouring caves containing depictions of several species in similar proportions, but otherwise there was no overall pattern.[103]

The importance of Leroi-Gourhan's work is enormous, not so much because of its results but through its impact on the field and on current and future research. It constituted a leap forward and, like all first attempts, it oversimplified, went too far too soon, and made many mistakes. As with any new and powerful theory, there was an irresistible temptation to bend ambiguous facts to fit, while ignoring others. There was an excessive tendency to assume that Palaeolithic artists thought like twentieth-century French structuralists. Leroi-Gourhan felt that he had established the 'blueprint' for cave-decoration, the framework on which a variety of symbols and practices had been hung through the period. His results are too erroneous to be fully accepted, but also too revelatory to be dismissed: some of his results, as we have seen, cannot be faulted. He found order and repeated 'associations', but not a universally applicable formula.

Inevitably, he modified his views over the years; Laming-Emperaire, however, who initiated this approach, eventually abandoned it in favour of a theory inspired by the work of Raphael and Lévi-Strauss, which virtually reverted to the totemism so beloved of early ethnographers, and which Reinach had presented (together with hunting magic) as an explanation of Palaeolithic art.[104] Breuil and Bégouën had opposed this view, since it meant that animals were mythical ancestors of human groups, and there would therefore have been taboos against killing or eating them – whereas they saw the depictions as killed animals.

Laming criticised the 'topographic determinism' which had arisen around the sexual dualism theory and which, as shown above, meant figures and signs were 'sexed' by their location, in a perfectly circular argument.[105] She maintained that the horse and bison were probably fundamental, but did not necessarily constitute a sexual opposition. Instead she wondered if they might be an expression of exogamy practices: ie a decorated cave was a general model of a group's social organisation, with animals of various sizes and ages representing either different generations or the mythic ancestors of certain clans. For example, a big bull and a horse at Lascaux would be clan ancestors. The male and female aurochs at Lascaux should not be seen as a bull and cow, but as two members of the bovid clan, such as a brother and sister. Two animals of different species and sex would be a matrimonial couple, while two of different species and the same sex would be an alliance.[106]

She believed that the main panels were origin myths, the principal animals were ancestors, and the secondary species were allied groups. The artists were not displaying a system or formula, but telling a story within a system. This approach was more flexible than that of Leroi-Gourhan because it did not expect everything to remain stable through time and space, but rather to be different in each cave. Unfortunately, Laming-Emperaire died before she could develop the idea further, although it is difficult to see how to do so without recourse to ethnographic analogies.

Whatever the faults and inadequacies of the work of these two scholars, they did succeed in changing irrevocably the way in which we think about Palaeolithic art. The images could no longer be seen as simple representations with an obvious and direct meaning; instead it was realized that they were full of conceptual ideas – they were the means by which the artists expressed some more metaphysical concept. Their work

Fig. 114 *Photomontage of the horse panel at Ekain (Guipúzcoa), c 4 m wide. Note the red bison at top right, and the 'M' mark on the horse hides. Probably Magdalenian. (JV)*

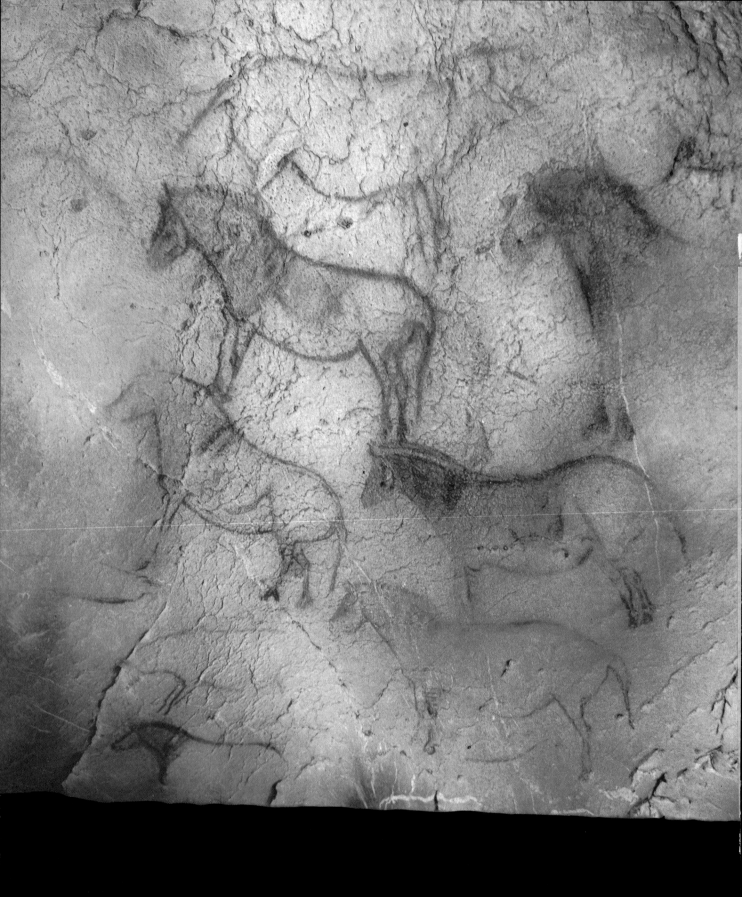

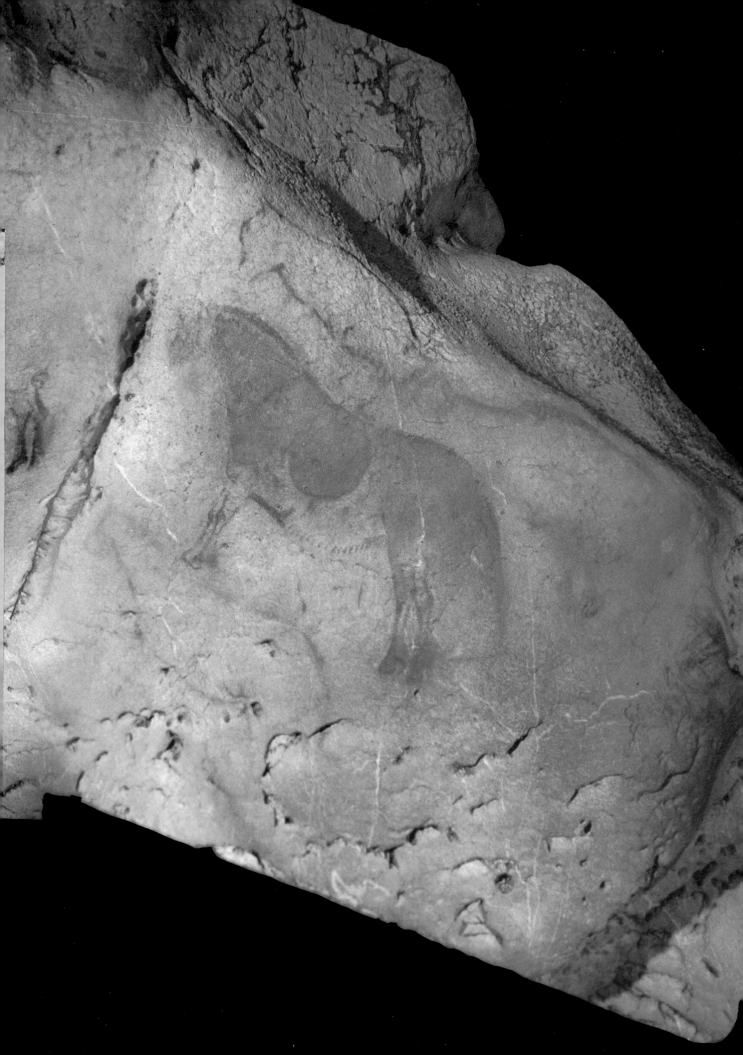

constitutes a very important transitional phase following the traditional approach of compilation and intuitive and ethnographic explanation. All current and future research will draw on their ideas and results, and modify or discard them. Breuil, Laming-Emperaire and Leroi-Gourhan would all have agreed that, for work to progress, we must not sit at their feet but stand on their shoulders.

It now remains, therefore, to look at a few of the varied directions in which current research is taking us.

Why is it where it is?

One of the most interesting new pieces of research has focused on the distribution of figures, and the shape of the wall beneath each one, in four neighbouring Cantabrian caves (Castillo, La Pasiega, Las Chimeneas, Las Monedas). It was found that in each of them the figures had been distributed according to the spatial characteristics of the cave, but there emerged clear differences between the content of 'concave' and 'convex' rock-surfaces: no less than 88.8% of horses were on concave surfaces, together with 93.1% of hinds, 87.7% of caprines, 100% of stags and 100% of hand stencils; on the other hand, 78.7% of bison and 82.8% of cattle occurred on convex surfaces.[107]

Moreover, the horse and bison were clearly the main representatives of the 'concave' and 'convex' groups respectively, and were usually bigger, more detailed and more complete than the rest of the animals. These 'hollow horses' and 'bulging bovines' confirm the existence of a horse/bison dualism and the importance of these species in the art.

Of course there are exceptions to the concave/convex rule, even in these caves – almost every statement made about Palaeolithic art can be countered with exceptions – but a preliminary look at other caves suggests that there may be a degree of consistency: for example, in Covalanas, the only 'convex' figure is the bovid, while all the hinds and the horse are 'concave'; at Altamira, all the bison on the ceiling are 'convex' whereas the deer and horse are concave; at Font de Gaume the bison and reindeer seem to be on convex surfaces; the Tuc d'Audoubert clay bison are on convex, the Cap Blanc sculptured horses and the Tito Bustillo painted horses on concave surfaces; while almost all the hand stencils of Gargas, like those of Castillo, are on concave surfaces.[108]

Much more work will have to be done before we can assess how widespread is this careful selection of surface-shape. Doubtless many figures and caves will not conform, because no scheme can ever totally fit an art-form spanning so many millennia. Nevertheless, this extension of the approach of Laming-Emperaire and Leroi-Gourhan may well prove to be a breakthrough.

A different observation about the position of art in the caves concerns its visibility and accessibility. It is not always easy to judge whether the difficulties we experience in reaching certain galleries existed in Palaeolithic times: for example, Montespan's art can be reached only by wading upstream, with water up to one's waist and, at one point, having to duck right under when the ceiling comes down to water-level; yet on very rare occasions, as in 1986, the cave is dry. Similarly, graffiti in the Tuc d'Audoubert (which now has to be entered by boat) show that it was visited in 1685 and 1702 by clerics in periods of great dryness.[109] One therefore cannot be sure whether the artists got their feet wet or not.

Visibility and access can also be affected by stalagmites and stalactites, and in some caves such as Cougnac these appear to have been broken by Palaeolithic people, thus making the figures more visible from a distance. One can divide cave-art into the clearly visible, the obscure, and the hidden – or, in other terms, the public and the private.[110] The Hall of the Bulls at Lascaux or the spotted-horse panel at Pech-Merle are good examples of visible art in large chambers, whereas there are many examples of art tucked away, either in inaccessible galleries (such as that of Fronsac, 35 cm wide) or in nooks and crannies of large chambers (eg the signs hidden in the folds of the 'lithophone' at Nerja).

It has been noticed that the parietal art is frequently associated with 'bouches d'ombre': ie chasms or entrances to lower galleries, many of which have water in them (eg at Tito Bustillo), and which may have been linked with myths concerning the underworld, the cult of the earth mother, and 'chthonic deities':[111] for example, at Ekain the farthest frieze of horses is located above two holes which lead to the dark and mysterious depths of the cave; in Travers de Janoye (Tarn) the parietal art is clearly arranged around a large fissure.[112]

In all regions and at all times, caves have been associated with placenta, with ideas relating to maternity and birth, and with the entrance to the underworld. Boundaries are areas of confusion to human beings and, like the orifices of the body, are sacred places full of taboo; caves certainly constitute a boundary between the outside world and the underworld, and thus have a special significance in many mythologies. Water plays a major role in virtually every known religion, and running water in particular is frequently looked upon as potent, dangerous and linked with spirits. It is therefore highly probable that where running water visibly crosses the boundary of the underworld, in either direction, it may take on special significance. This effect would be heightened where water enters the earth in an impressive setting (Mas d'Azil, Labastide), or where it emerges as a hot spring.

It has been suggested that these factors are likely to have played a role in whatever beliefs and rituals lie behind Palaeolithic parietal art – what might therefore be called the 'rites of springs': is it a coincidence, for example, that large clay models have been found in only two caves (Tuc d'Audoubert, Montespan) through which rivers flow, and that both contained a headless snake skeleton? A large number of decorated caves are located very close to springs, and there is a marked correspondence between certain parietal sites and thermal/mineral springs:[113] for example, the four caves in the Monte Castillo overlook the village of Puente Viesgo, through which a geological fault runs, responsible for a number of hot springs which have been found suitable for medicinal purposes; it would be unsurprising if in the last Ice Age very special significance was accorded to springs which did not freeze and/or which were thought to have medicinal powers.

Recent work, therefore, has focused on the distribution of art within caves, and on that of particular decorated caves within the regions where they exist. Little attention has been paid to the problem of why the distribution of parietal art is so restricted in Europe, unlike that of portable art and despite the existence of perfectly good caves in Germany, for example.

One theory is that southern France and northern Spain constituted 'refuges' into which people migrated in waves from other parts of Europe during periods of climatic deterioration from about 25,000 years ago onward: the 'cave-art regions' were more temperate, and had a greater abundance and diversity of fauna available, and people were able to live there by exploiting the plentiful reindeer and Atlantic salmon.[114] Salmon-fishing would have led to increased sedentism, and thus a greater need for elaborate procedures for resolving conflicts and marking territory: hence the cave-art sites were established as permanent ceremonial locations; while the portable art was perhaps linked to the more mobile life and seasonal aggregations associated with reindeer exploitation. In short, the art is seen as a manifestation of different social responses to processes of climatic and economic change and population movement.

Unfortunately, the theory does not stand up to scrutiny, simply because parietal art, as we have seen, is not limited to Franco-Cantabria, merely concentrated there; in any case, reindeer were not the only staple resource in the Upper Palaeolithic – many sites, depending on their location and surrounding topography, had an economy based on horse, bison, red deer, ibex, or a mixture of several species. In addition, fish began to be exploited on a significant and systematic scale only at the very end of the period, long after the main phases of art production;[115] and, of course, some pockets of parietal art such as the Rhône valley were beyond the range of the Atlantic salmon.

Consequently fish and reindeer exploitation can have had little or no bearing on the production of Palaeolithic art. It is still possible that France, Spain and the Mediterranean coast represented a refuge at times. However, we are unlikely ever to discover precisely why this art flourished where and when it did – it may constitute a kind of long-term Palaeolithic equivalent of Classical Greece or Renaissance Italy.

How many artists were there?

But was Palaeolithic art produced in a steady stream over the millennia, or does it represent a large but sporadic output by a relatively small number of gifted artists during this long period? Some scholars still hold the former view, but most now prefer the latter.[116]

As mentioned earlier (p.105), experiments suggest that many parietal and portable figures could each have been executed in a very short time by an artist with a modicum of talent and experience: all the black-outline animals of Pech-Merle could have been done in less than three hours, and probably represent a single artistic episode: indeed, this cave is thought to have had no more than three or four such episodes – for example, analysis shows that its six black hand stencils were done with the same pigment and used the same hand.[117] Together with the archaeological and pollen evidence, these facts suggest that, far from being a great tribal sanctuary, Pech-Merle was visited and used only very rarely.

Similarities between Palaeolithic images – primarily in portable art – have often led, in the past, to theories about 'schools' or workshops: Limeuil, for example, was seen as a centre of artistic and magic education. Images on stones and bones were thought to be artists' notebooks or sketches, and the 'mediocre' specimens among the masterpieces were the work of pupils. While such notions are somewhat fanciful and owe much to nineteenth-century views of art, some modern scholars still attempt to estimate and compare the percentages of 'masterpieces' and 'mediocre pictures' in different sites,[118] an exercise which is entirely subjective and takes no account of the artist's intentions.

Some clusters of images are stylistically and technically so similar that it is virtually certain that one artist or group is responsible. Hence, at La Marche, a particular style, technique and set of conventions (and errors!) have been recognised by those most familiar with its engravings.[119] Few would doubt that the Mas d'Azil and Bédeilhac fawn/bird spear-throwers (p.82) were by the same artist (or, at least, two artists, one of whom had studied the other's work); the same could be said of the portable and engraved hind heads of Altamira and Castillo (p.58), some of the human profiles of La Marche and neighbouring Angles-sur-l'Anglin (p.58), the Labastide isard heads (p.82) and certain portable bison-head engravings from Isturitz.[120]

On the other hand, other collections of portable art seem to be by a number of different artists: for example, the Castillo engraved shoulder-blades, or the slabs at Gönnersdorf where analysis of details suggests that at least a dozen different artists were responsible for the mammoths.[121]

The search for individual artists is by no means a new phenomenon in Palaeolithic art: de Sautuola himself saw the Altamira ceiling as a unified work, and in the Cartailhac/Breuil monograph on the cave, there was some discussion of 'authors', and speculation that not all the cave's pictures were by the same hand. Until the end of his life Breuil insisted that one artist of genius could have done all the polychromes of Altamira.[122] Leroi-Gourhan, on the other hand, refused to accept that one could attribute even two adjacent images to a single artist.[123] He was obviously correct, in so far as it cannot be proved that more than one picture was done by a particular person; but there is a place here for intuitive reasoning: for example, Pales was certain that at least three of the ibex in Niaux's Salon Noir were by the same artist.[124]

One scholar, Juan-María Apellániz, has for some years been trying to establish firm criteria for recognising the work of individual artists, to transform what have previously been intuitions into a reasoned method. He looks for an original way of drawing, a particular '*tour de main*', the repetition of idiosyncrasies, peculiarities and details of technique and execution. There are variations, of course, since artists do not repeat themselves exactly, and may well have changed style slightly through the years: the same artist could certainly produce two very different figures ten years apart, while, as we have seen, a change of tool can also affect the hand. Only complete figures can be used.

It is advisable to work directly from the original images, rather than from tracings and copies by other people which, as we have seen (chapter 3), can exaggerate the similarities or differences between figures. Occasionally Apellániz has attempted to work from other people's tracings,[125] but more often uses direct observation, sometimes backed up with photographs and copies, as in the case of the Altamira ceiling, where he confirmed earlier scholars' intuitions that the ceiling was largely the work of one 'master'.[126] It is relatively easy to compare the standing and curled-up bison figures, since, as mentioned earlier (p.119), they share many characteristics; however, it is much harder to extend the comparison to animals of different species such as the horse and hind on the ceiling.

Studies of this type can be particularly easy on portable objects, where similar figures often form friezes and were almost certainly the work of one artist (eg on the Torre bird bone, Fig. 42, all the horns have their ends drawn in the same way, probably by the same hand); the same applies to objects such as bone discs with decoration on both sides: on that from the Mas d'Azil with a calf on one side and a grown bovid on the other, or that from Laugerie-Basse with a chamois on each side, the two figures are identical in their use of space, their proportions, composition, silhouette and technique.[127]

In his work on parietal art, Apellániz has produced mixed results. At Altxerri and Ekain, he has 'identified' different artists or groups working in different parts of the caves.[128] At Las Chimeneas he believes that all the black stags are probably by the same hand, and at Lascaux the same applies to the four huge black bulls, whereas the cows seem to have a number of authors.[129] One of the clearest sets of similar figures is found in a group of caves: in Covalanas and, 150 m away, La Haza; in Arenaza, 45 km to the east, and La Pasiega, 60 km to the west. The fidelity to the 'general canon' of composition in hind and horse figures, the dotted outlines, the similarity in details such as eyes and extremities of limbs are such that he claims this group of caves can be seen as a 'school': ie the 'master' or principal artist of Covalanas had disciples or companions who worked in the other caves.[130] However, he does not believe an artist worked outside a particular cave: each cave may have formed part of a 'school' of this type.

How valid is this approach? The criteria presented as objective and reasoned are still rather subjective and intuitive; he is assessing the degree of similarity between figures which may be by one hand, but which may also simply display the accepted canons and conventions of a particular period and culture.[131]

In an effort to counteract these problems, Apellániz is developing statistical techniques, using a series of variables and measurements of various parts of the animal outlines (like those made by other scholars on bison figures, p.122), and subjecting them to factor analysis in order to assess their degree of similarity: so far this method has confirmed that the Chimeneas stags, for example, are extremely similar in the treatment of the neck, back and rump.[132] It will never be possible to prove that the same hand was responsible, but this approach has certainly established a very high degree of probability in some cases.

We can only speculate about how Palaeolithic art's 'rules' and 'canons' were passed through space and time so faithfully. It is likely that most decorated caves were visited only rarely in the Upper Palaeolithic, and in any case many of their figures are hidden. Portable art, however, was probably circulating over great distances, and people (including artists, presumably) were often moving seasonally, and coming together annually or periodically in aggregation sites when much could be taught and learnt.[133] Finally, there must have been a great deal of art in perishable materials and, as we have seen, there were also 'permanent' images in the open air. It seems certain, therefore, that Palaeolithic people were surrounded by and familiar with many forms of art, and so there is really no mystery about how its styles and techniques were passed down.

Art as information, art for survival

Indeed, it is almost certain that some of the art itself constituted a means of storing and transmitting information of various kinds: this is particularly likely for some non-figurative motifs. Once again, such notions are by no means new: Lartet interpreted some sets of marks on bone and antler as 'hunting tallies' and, as mentioned earlier, other motifs were thought to be artists' signatures or proto-writing.

A different kind of role for parietal non-figurative marks was first suggested by Breuil in 1911: he saw that some panels at Niaux, covered in dots and lines, were located just where the main passage divided, and felt that the marks might therefore be topographic guides. There do seem to be occasional links between 'signs' and important places in a cave: at Lascaux, for example, some lines of dots are located at points of topographic transition, and similarly in other sites signs are positioned where passages turn, become narrow or branch off.[134]

In an extension of this view, and of the similar idea that certain caves might have been decorated as symbolic animal drives/pounds (see p.155), some scholars have even suggested that the artists took advantage of the topography of cave-walls to provide or store information about the surrounding landscape: for example, Leroi-Gourhan speculated that horse and bison were primarily in the spacious, central parts of caves as these represented the wide-open pastures, while figures of felines clustered in the dark and narrow depths because these represented their lairs.[135] Others believe that some cave-art might be a model of the local Palaeolithic hunting territory: it has been pointed out that the axes of the main cave of Ekain exactly parallel the line of flow of the streams in the area, while a conical rock in the cave is like the hill housing the cave itself.[136] It is an interesting idea, but very difficult to put forward in more than a handful of cases, particularly as we do not know how certain features of the landscape (such as stream-flow) may have changed. Nevertheless, it is quite possible that some of the art acted as a mnemonic device in this way, to instruct the younger members of the group (see below).

A few items of portable art have been interpreted as actual maps: a stone from Limeuil with engraved meanders on it was thought to be a possible river map; however, the best known is a fragment of mammoth tusk from Mezhirich (Ukraine), dating to the twelfth millennium BC, which is engraved with narrow bands, transverse lines and some geometric forms in

Fig. 115 *Tracing of the so-called engraved 'map' on the ivory plaque from Mezhirich (USSR). (After Gladkih et al)*

0   1   2 cm

the middle. It has been claimed to represent a map of a river with four dwellings along it and some fishing nets across it;[137] but in fact the object was published upside down in terms of the direction in which the marks were made, and the central geometric forms are totally unlike the Palaeolithic huts of this region – in any case, the engraving appears to have been accumulated rather than produced as a single composition.[138]

The 'hunting tally' theory about sets of marks has been revised by a number of scholars in recent years: some have used ethnographic analogy to interpret groups of dots and lines on bones as being connected to games of chance for telling fortunes;[139] others prefer to search for evidence of Palaeolithic proto-mathematics. Absolon used to look for a system based on 5 and 10, but since this is now considered to be imposing our modern norms on the past, other scholars such as Boris Frolov claim to have found repeated multiples of almost every number from 1 to 10 on portable objects from the USSR and elsewhere.[140] There is said to be a particular frequency of 5 and 7, with 3 and 4 as 'supplementary number sets'. Frolov sees this numbering system as the basis of an incipient Palaeolithic science.

He has been vigorously opposed by Alexander Marshack, who points out[141] that these numbers are quite random, and the discovery of any repetitions in Palaeolithic engravings is more indicative of a researcher's ability to count than of any number-system in the Palaeolithic, and that claims about isolated sets of marks like these cannot be tested. There is no reason to see them as numbers (let alone arithmetic) when, in Marshack's view, they are more likely to be 'notations' – ie sets of marks accumulated as a sequence, perhaps from observation of astronomical regularities. He has tested linear sets of such marks (most notably a serpentine set on an Aurignacian object from abri Blanchard) for 'internal periodicities and regularities', particularly as possible examples of lunar notation, with varying degrees of success,[142] though he believes that there are also non-lunar notations.

Whatever the validity of their results, these studies have at least focused attention on a type of marking which previously was ignored or dismissed as random; it is now clear that they are coherent and ordered, and were carefully made over a period of time. Similarly, a recent study of the painted pebbles from the Azilian (the very end of the Ice Age) revealed that the 16 signs drawn on them were found in only 41 of the 246 possible binary combinations (indicating that some sort of 'syntax' was employed) and that there was a predominance of groups numbering from 21 to 29 or their multiples, which may again have some connection with lunar phases or lunations.[143]

The phases of the moon would certainly have been the principal means available to Palaeolithic people for measuring the passage of time, and other scholars have likewise sought evidence of lunar observation, but of a different kind: an ivory plaquette from Mal'ta (Siberia) has hundreds of pits engraved on it in spirals; these have long been interpreted as symbols of moon worship. Frolov claims that there are 7 spirals, the biggest of which has 243 pits in 7 turns, and the others 122 (243 + 122 = 365). By analogy with similar motifs found among modern Siberians, he interprets it as calendrical ornamentation (243 days being the gestation period of the reindeer, the staple food at Mal'ta, as well as the length of the winter, while the summer lasts about 122 days).[144]

A different method of noting the passing of time, of course, is by means of the seasons: those who uphold the 'hunting art' theory point out that there are very few 'spring scenes' (emaciated animals, pregnancies, births) and a preponderance of autumn images (fat animals, bouts, pre-copulatory behaviour), while Marshack has interpreted many groups of figures, especially in portable art, as representing particular seasons – though some of his interpretations such as the 'moulting animals' of La Colombière or Lascaux are by no means definite.[145] We have already seen (chapter 6) that little reliable zoological or ethological information can be extracted from Palaeolithic images, though a few details such as the hook on a salmon jaw are certainly indicative of a specific time of year.

Marshack has developed the concept of the 'time-factored' symbol to explain the periodic use of images or accumulations of notation. They were images of seasons, relevant to economic, ritual or social life, and made to be used at the correct time and place. The distinctive features of a season were depicted (eg on the Montgaudier baton) to provide a reference point in the calendar. Hence, where previous scholars saw the 'penniforms' around the 'Chinese horse' at Lascaux as missiles or phallic symbols (depending on their pet theory), Marshack would see this as a horse in summer coat running among plants or ferns.

Marshack's importance lies not only in devising new techniques for investigating and documenting the art (see chapter 3 and Appendix), but also in asking new questions about it, and in postulating that many images were built up and re-used over long periods. He believes that we are faced with a very complex tradition of symbolic accumulations which were regionally specialised and differentiated. The notations are merely one type of sign system: many other 'marking strategies' were in use simultaneously. He has focused on what seem to be non-figurative, geometric markings found on non-utilitarian artefacts, always studying

Fig. 116 *The enigmatic 'spade' sign at Le Portel (Ariège) with a painted horsehead nearby. The man painted around a stalagmite [see Fig. 52] is below the sign. Probably Magdalenian. The sign is 45 cm across. (JV)*

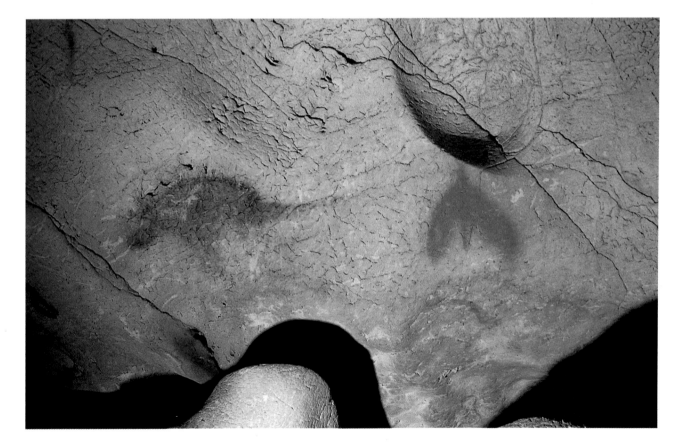

the sequence and process of accumulation, rather than just the finished images.

In this way, he has found motifs of different types recurring throughout the Upper Palaeolithic, such as zigzags, 'ladders', double-lines, and (particularly in the Soviet Union) fish-scale or netlike patterns. One of the most fundamental is the 'band motif' on portable objects, which may be equivalent to the finger-tracings on cave-walls and which, he has speculated, may represent one aspect of a water-related symbolism[146] – whereas, predictably, Breuil saw the parietal 'meanders' as serpents, and Leroi-Gourhan saw them as perhaps phallic. Marshack does not claim that they actually represent water, but that they were 'iconographic acts of participation in which a water symbolism or mythology played a part', a theory which fits well with the possible connection, mentioned above, between certain caves and water entering or coming out of the earth.

These marks are no longer seen as random accumulations but instead as a sophisticated, sequential, additive image system; no longer dismissed as idle doodlings, but considered to be possibly the most complex element in Palaeolithic iconography. The act of image-making was the important factor, and it was not always done in a public or participatory setting: as we have seen, a great deal of parietal art is hidden away in inaccessible recesses and tiny chambers, such as the horse and 84 'P' signs in a narrow passage of the Tuc d'Audoubert (see Appendix), or the little chamber at La Pileta with its animals and double fingermarks,[147] or indeed the tiny engraved 'sanctuary' in Karlie-ngoinpool, Australia (see p.32). It is assumed that different people squeezed into these crannies at different times to perform some symbolic activity.

Some images, both portable and parietal, both hidden and public, were clearly 'touched up' at times. Opinions still differ about what kind of behaviour lies behind this phenomenon. Marshack firmly believes that these images were used and re-used, and has published many examples such as engraved animals on pebbles and bones from Polesini, Italy, which had legs or muzzles added at some point after they were originally drawn.[148] As we have seen (pp.75-6), some of his claims that each of a series of marks was made by a different point or hand are simply not verifiable – a variety of lines or marks could have been done by the same tool or hand, and one cannot take this as proof of long-term retouching.[149]

Many theories have previously been put forward to explain this phenomenon which seems to be most frequent in the Magdalenian of Franco-Cantabria. A proponent of hunting magic such as Bégouën saw it as the renewal of an image which had 'worked' before, the act of drawing thus being more important than the finished picture (conversely, the images would be broken if results did not comply with the prayers). Other scholars believe that some figures, as at Lascaux, were simply improved or rectified over a short period (they show a unity of style), the artist experimenting until satisfied with the figure's position and quality; while others suggest that the multiple hoofs, muzzles, etc (eg on reindeer at Ste Eulalie, Lot) may be an attempt to depict movement.[150]

The fact that we cannot assess the timespan involved prevents us from establishing which explanation is correct; however, where carvings are concerned, Marshack's view seems more plausible, since some of them – such as those of Vogelherd, or even the Tata plaque – are worn and polished from long handling, and the Vogelherd figurines seem to have had marks added to them at times.[151]

Fig. 117 *Tracing of some painted and engraved horses in the Panneau de l'Empreinte at Lascaux (Dordogne), showing extra eyes, ears, muzzles and legs. Probably Magdalenian. The 'multiple' horse covered in arrowlike marks is about 1 m in length. (After Glory)*

Even though some marking activity may have been private, it is nevertheless true that much of it is on open view, both in caves and on objects, and must therefore contain messages of various types.

Interpretations of Palaeolithic art have been governed largely by the knowledge and preoccupations of academics in different decades. Just as 'art for art's sake' was the theory which best suited the end of the last century, and 'sympathetic magic' was taken from the ethnography of the early years of this century, so it was no coincidence that the structural and sexual interpretation arose during the 1950s and 1960s (the era of the sexual revolution, and of Lévi-Strauss) or that ideas about astronomical observation were developed during the Space Age.

As with Stonehenge, each generation re-invents the evidence in its own image. Just as our computer age produced grossly exaggerated views of that monument as an accurate eclipse predictor, the new technology has inevitably led to some novel ideas about the purpose and meaning of the earliest art. A great deal of research has been done in recent years into information and its processing, and into the nature of memory and how we can use our memory centres more efficiently. One of the major aspects of this work concerns how art can be employed to transmit information and imprint it on the mind.

One such approach has been an attempt to apply 'Information Theory' to cave-art; it was thought a useful technique since it ignores the potential meaning of the figures, and concentrates solely on their 'quantity of information'. First applied to language in 1945, the theory defines a message's quantity of information as a function of the number of letters of which it is formed and the language in which it is written. The unit of information is called a 'bit' (an abbreviation of 'binary digit').

As we have seen, it is a fair assumption that in the last Ice Age most of the cave-art contained a 'message' which was not aimed at us and which we cannot understand clearly. This should not matter to Information Theory, but there is a more crucial obstacle in the fact that we do not know what kind of classificatory system should be employed in our analysis. This is important, because the frequency of each 'symbol' directly affects its quantity of information; the more examples of a particular motif in a sample, the fewer 'bits' are allotted to each of them. But how should we divide up the figures? Should bison be separated from aurochs, or stags from hinds?

In short, the applicability of Information Theory to cave-art is severely limited, and the few results achieved so far could have been produced equally well by common sense:[152] it is obvious that in a panel of similar figures, such as the horses of Ekain (Fig.114) or the bison-ceiling of Altamira (Fig.7), each individual figure is of less information value than a single, unique figure, such as the bizarre 'sign' in the cave of Le Portel (Fig.116), which resembles a 'spade' from a playing card and which has also been seen as a hut, a vulva or a bird in flight. But it does not mean that we are any closer to understanding what that information might be!

Exercises of this type are akin to a Martian coming to Earth and trying to calculate the information contained in an incomprehensible English message by lumping it together with some Polish, Basque and Latin in order to establish the probabilities of the different symbols used. They involve an attempt to measure the degree of complexity of a structure or a message, but do not take into account the context; and, as we have seen, with cave-art, the context may be even more important than the content.

In this computer age, it was perhaps inevitable that a cave with figures on its walls should be interpreted as a great 'floppy disk', a storehouse of information for long-term retention and recall. Such an explanation does have the merit of taking into account the position and distribution of figures within a cave, and sees the blank spaces between the groups as part of the message. Indeed, the art itself is only part of the experience: the task of reaching it, and the strange sensations and occasional dangers involved, are all part of the process of imprinting the message indelibly. It has long been assumed that cave-art, at least, has much to do with the initiation of

Fig. 118   *Photomontage of the entire 'shaft scene' in Lascaux (Dordogne), showing rhinoceros, dots, apparently wounded bison, bird-headed ithyphallic man and 'bird-on-stick'. Probably Magdalenian. Total width: c 2.75 m. (JV)*

young members of the group, and the 'information' and 'survival' theories have again brought this to the fore.

We have seen (p.110) that rock-art was done out of doors in the same period, but caves are where parietal art has survived most successfully, and the secrecy and darkness involved in a visit to them must have been a memorable experience for any initiate. Quite apart from the physical difficulties encountered in some caves, whether rock-formations, flowing water or holes to crawl through, there is the utter blackness, the total silence, the loss of sense of direction in the often labyrinthine passages, the change of temperature, and the frequent sense of claustrophobia.[153]

The fear of being abandoned, lost and alone in the dark, would have concentrated the mind wonderfully and prepared the apprehensive initiate for anything. Similar techniques are still in use today for more sinister purposes, such as brainwashing or debriefing: victims are taken to alien, unpleasant or unfamiliar surroundings, subjected to physical and mental discomfort, made confused and uncertain about what is to happen; their entire frame of reference is removed, the everyday world is completely undermined – a phenomenon which Henry Miller called 'moving day for the soul'. The technique of sensory deprivation also concentrates the mind, and a cave would certainly affect all the senses of scared initiates quite radically and leave them vulnerable to indoctrination.

The advocates of this idea do not assume that Ice Age people were aware of all these factors – no doubt they used instinct as much as planning – but the experience of centuries will have taught them which techniques were most successful; and success was crucial since the tribe's very survival would depend on its accumulated learning and rules being transmitted and imprinted effectively and indelibly. Hence the careful positioning of certain figures in niches, or in places where they could be effectively and suddenly revealed by cunning lighting – these were the rudimentary 'special effects' of the time which must have had considerable impact on impressionable and expectant minds. One can speculate that the mass of unsexed human figures in the dark cave-depths represent these adolescents and their ambiguous, transitional status.

But how exactly could these images transmit information? Clearly, we cannot know precisely what the different animals or signs were 'metaphors' or mnemonics for, which rules or events they were connected with. But they must have been vivid, overwhelming visual impressions, especially by the flickering flames of small lamps which, as mentioned earlier (p.110) make the animals seem to move; and pictures of this sort are extremely powerful conveyors of information: the cliché that 'a picture is worth a thousand words' is all too true here, and cave pictures still shout down the millennia, though now their language is alien to us.[154]

We know from modern studies that the human mind can 'take in' up to half a dozen items at a glance, and these convey far more information to the mind when arranged in familiar patterns: for example, if one glanced at the above cliché, and at seven unconnected words, the cliché would be perceived more clearly and remembered. The same is probably true of Palaeolithic art, which certainly comprises a 'vocabulary' of symbols, some of which must have had considerable information value, and certain combinations of which may have had special significance.

One further idea concerning mnemonics involves not the thinking of the computer age, but methods developed by the Ancient Greeks and Romans, and maintained in medieval Europe. These entailed memorising vast quantities of information through a technique of impressing 'places' and 'images' on the memory. Before the invention of printing, a trained memory was often of vital importance; people would therefore imagine, in their 'mind's eye', a large deserted place such as a great cathedral or a whole series of rooms. By walking around it mentally, and memorising thousands of places, each associated with something they wished to remember, they could later repeat the 'journey' and recall everything perfectly. The Greek poet Simonides, who wrote the first text about this method, called it 'inner writing'; and some Roman scholars used it to commit long texts to memory – Cicero's speeches, legal points, or Virgil's entire *Aeneid* (9,700 lines, which could even be recited in reverse order!).[155]

Committing information to memory in this way is second nature to Australian Aborigines who use vast stretches of desert, containing thousands of special places (water-holes, hollows in trees, etc) which they can associate with the Dreamtime wanderings and adventures of their ancestors. In this way, a huge amount of tribal lore, vital to survival, is preserved and imparted little by little to the next generation during initiations and festivals. Art plays a crucial role here; and it is quite possible that similar methods were employed during the Upper Palaeolithic period, both in the open air and in the caves. The latter are suitably quiet, deserted settings for a 'mental walk', containing numerous special places such as niches and concretions by means of which information could readily be committed to memory.

In short, the concept of computerised information has combined with the ancient art of mnemonics to produce a new view of the earliest art, worthy of the 1980s. Naturally, little evidence survives to support the idea, apart from the obvious fact that some of the art is in caves, and carefully set out within them. We have no proof of Palaeolithic initiation ceremonies, nor of how information was transmitted or imprinted.

Some scholars believe that parietal art constitutes a system of communicating information, though we need to learn much more before we can feel confident that it is a semiological system capable of transmitting coded messages.[156] Others claim that portable objects are more likely than parietal art to be involved in the storage and rapid transmission of information, for the simple reason that they are portable and not all of them are decorated. Certainly their different motifs do display interesting distribution patterns, as, for example, in the Magdalenian of Cantabria, where there is distinct regional variation: some motifs are widespread, while others are unique to major centres ('aggregation sites') such as Altamira, or to particular areas, and may thus mark group boundaries.[157]

It is believed that having a set of motifs or a common art-style, different from those of one's neighbours, helps to maintain a group's identity and cohesion (an adherence to norms is the basis of a culture's stability), while motifs shared with neighbours would foster intergroup solidarity[158] – a 'portable art version' of Leroi-Gourhan's view of complex signs as ethnic markers; it has even been claimed that some objects such as 'Venus figurines' were display items indicative of a system of communication and information-exchange, designed to maintain contact and alliances between the participants in mating networks which had become strained through climatic deterioration and expansion of hunting territories![159]

Unfortunately, there is no evidence whatsoever to support these fanciful ideas – we have seen, for example, that 'Venus figurines' are by no means a homogeneous group limited to two short periods, and thus there is nothing to link them with 'alliance networks', while the environmental deteriorations thought to have triggered this crisis (and the above-mentioned migrations of people into Franco-Cantabria) have been assumed, not demonstrated. There were many climatic fluctuations in the course of the Upper Palaeolithic.

Like all the other views outlined above, these theories merely reflect one of archaeology's current preoccupations – prehistoric social systems, the delineation of group boundaries, and the links between style and information content; and, like the preceding theories, they may well contain some truth. Time will tell, but meanwhile at least one of the proponents has decided that the 'art as adaptive information' idea is really a truism and based on some very weak assumptions[160] – and in any case should not be assumed to apply to all of Palaeolithic art. Other recent ideas are simply too vague to be useful: for example, the view has been expressed

that group cooperation was required for large-scale communal hunts
(another truism), and that cave-art may represent the rituals involved in
reinforcing the authority of the hunt coordinators; alas, there was no
suggestion as to how this was done.[161]

All in the mind

Finally, the most recent approach to Palaeolithic art has focused on the
origin of the oldest motifs, both in Europe and Australia, and drawn on
current research in neuropsychology. 'Phosphenes' are subconscious
images, geometric shapes that seem to be present in the neural system and
visual cortex of all human beings: some of the earliest art is thus thought to
be an externalisation of these images. This would explain the similarity
between the finger tracings in Europe and Australia – they sprang from a
common neural circuitry.[162]

These geometric forms are entoptic (ie can be seen by the eyes when the
eyelids are shut) and are experienced, for example, during migraine
attacks. One way of inducing them is to alter the state of one's
consciousness by trance or hallucination: psychologists have shown that
the early stages of trance can produce such hallucinations in geometric
forms. Hence, it has been argued that many simple Palaeolithic 'signs' are
shapes of this type,[163] although any collection of non-figurative art is likely
to include lots of marks which resemble entoptic phenomena, simply
because there are very few basic shapes that one can draw, whatever the
motivation or source: dots, lines, squiggles and basic geometric forms.[164]

As we have seen, there is some likelihood that shamanism played a
major role in the production of Palaeolithic art, and it has recently been
claimed, largely through analogy with well-documented shamanistic art in
southern Africa, that, in addition to the 'signs', many animal figures and,
especially, the composites in Palaeolithic art may be hallucinatory images
– ie they represent the insights and experiences of shamans in trance
performances, and should thus be 'read' metaphorically.[165]

This view may well account for some of the more disturbing cross-
cultural parallels in images that are far more complex than meanders or
simple geometric shapes. For example, in southern African rock-art, there
are images of a medicine man in a trance next to a dying eland (an analogy
between the animal's state and the shaman who 'died' in trance[166]) which
resemble the shaft scene of Lascaux with its bird-headed man and
wounded bison (Fig.118); similarly, the Lascaux man, with his erection
and his 'bird on a stick', is remarkably like an Arizona rock-engraving
depicting an ithyphallic man with a bird on a stick.[167]

And what are we to make of the fawn/bird spear-thrower of Mas d'Azil
(p.83) which is identical (apart from the turd!) to a pose struck by Disney's
Bambi in a film made before the object was discovered? There is also a
striking analogy between one Palaeolithic scene and a Greek amphora of
the eighth century BC: the Magdalenian baton from Lortet shows a fish that
seems to be leaping up towards the genitals of a deer, and above are two
lozenges with dots inside; the amphora has a very similar composition,
except that the deer is replaced by a horse, and there are three lozenges.[168]

Are these simply remarkable coincidences? Or are they indicative of
some widespread phenomenon such as 'universal myths' of very remote
antiquity, or perhaps something akin to the phosphenes – images seen
under trance conditions? Future research may help us decide.

8
Conclusion

Readers may by now be wondering what exactly we do know about Palaeolithic art – and the answer is: both a great deal and virtually nothing. For a start, although we have about 275 decorated sites in Europe alone, and thousands of pieces of portable art, they probably represent a tiny fraction of what existed originally – most outdoor art has gone, so has anything in perishable materials, and we do not know how many decorated caves remain undiscovered. Even in the caves we have, untold quantities of figures may have been weathered or washed away, trampled into the floor or covered by stalagmite and clay – in rich sites such as Castillo or Trois Frères new figures are still being discovered today.

There is a vast literature on the subject, and hundreds of tracings of figures; but, as we have seen, many tracings, cave-plans and studies are inaccurate, incomplete or biased, and everything needs to be checked, improved and amplified. Much portable art lies unknown and unstudied in museums and private collections, while some major parietal sites still await adequate publication.

It is only with full and accurate facts at their disposal that prehistorians can hope to draw valid conclusions – but, as shown in chapter 6, it can be difficult enough to identify the figures, let alone assess what they are doing or what they might mean; indeed, some would argue that we should abandon the attempt to extract zoological or ethological information from the art, whereas others see these topics as crucial to our understanding of the art's message.

Thanks in large measure to the work of Léon Pales, rigorous and objective anatomical criteria have been applied to a large number of figures, and many doubtful cases have been weeded out. Currently much effort is being devoted to distinguishing between secure identification and mere interpretation, between what is definite and what is only probable: in future, prehistorians working on Palaeolithic depictions will have to be far more careful about presenting probabilities as certainties, or mere possibilities as probabilities.

As we have seen, the current emphasis is on total recording of the art: not just all the marks present but also their context. In a cave this entails study of its topography, sediments, climate, archaeological contents, and its use between the Palaeolithic and the art's discovery. Moreover, not only the art's location, date and subject-matter are investigated but also the technique of execution, and the time taken to produce it. Such topics may seem less fascinating than the art's meaning, but they are fundamental to any theorising, and it is no coincidence that Breuil devoted more attention to the question of 'when', and Leroi-Gourhan to 'how', in the latter part of their lives.

We need these solid facts. As Sherlock Holmes said, in *A Scandal in Bohemia*, 'it is a capital mistake to theorise before one has data. Insensibly, one begins to wish facts to suit theories, instead of theories to suit facts.' On the other hand, facts do not speak for themselves: at some point you have to choose a theory and test it against the data. Unfortunately, this has been done badly in the past: as we have seen, all-embracing theories were presented, and facts bent or ignored in order to make them fit. It is unlikely that anyone well acquainted with Palaeolithic art will ever try this again.

The reason is simple: we are faced here with two-thirds of known art-history, covering 25 millennia and a vast area of the world (it is ironic that so many books on art begin with a token picture of Lascaux, bearing in mind that it stands at the half-way mark in the history of art! – indeed, probably closer to us than half-way, since Palaeolithic art makes sense only if it was preceded by a long period of development on perishable materials). What we are trying to investigate ranges from simple scratches on bones and stones to sophisticated statuettes, and from figures on blocks and rocks in the open air to complex signs hidden in the inaccessible crannies of deep caverns. Almost every basic artistic technique is represented, together with an often simultaneous use of different conventions such as realism, schematisation, stylisation and abstraction. Most scholars would now agree that it is futile to try and encompass all of this within one theory.

Not all of it is necessarily mysterious or religious, though some parietal art is almost certainly tied to ritual and ceremony. Some is private, some public; some figures are discreet while others draw attention to themselves.

There is a certain thematic unity – profile views of the same restricted range of animals are present throughout in some areas, together with a mass of apparently non-figurative motifs – but, beyond this, any homogeneity is more assumed than real. As we have seen, the most recent studies emphasise that every cave has its own unique symbolic construction, and it is becoming ever harder to establish regularities in the midst of such wide variation.

Similarly, portable art is marked by great regional variation, both in forms and in motifs; it is a parallel, perhaps complementary art-form, reflecting different preoccupations, and it is extremely rare (as with the Altamira/Castillo engravings) that one artist seems to have tackled both.

One can therefore study the complex phenomenon of Palaeolithic art at three principal levels: the individual sites (either a cave, an object, or a group of objects); regional groupings (of motifs, complex signs, dominant animal species, styles, etc); and as a whole, seeking homogeneity and regularity. As the last option loses favour, efforts are being focused on the other two levels of study, as well as on how things changed through the period. As our data base becomes more complete and objective, it will become easier to identify regional and chronological groupings and characteristics, and hence theorise more securely about social systems and information exchange. In the past, we have tried to 'run' before we can 'walk', trying to extract more information than was available, simply because we dislike unintelligible things. But Palaeolithic art is perverse – every new 'piece of the jigsaw' raises fresh questions and casts doubt on the conclusions already reached. There are no absolute rules, there are always exceptions.

We have seen that the currently fashionable theories involve a cosmological structure, the marking of group boundaries, the fostering of cooperation, the recording of seasonal phenomena, the maintenance of mating networks, the initiation of the young, and shamanistic trance images. The more traditional theories include art-as-pleasure, hunting and fertility magic, totemism, and a macho preoccupation with hunting and girls. In view of the timespan and vast area through which Palaeolithic art is spread, it is probable that all these theories contain some truth, and that there were many other motivations which may never be known.

It is generally agreed that Palaeolithic art contains messages, no doubt of many kinds (signatures, ownership, warnings, exhortations, demarcations, commemorations, narratives, myths and metaphors); their basic function was probably to affect the knowledge or the behaviour of those who could read them. We, alas, do not know how to read them. We can analyse their content, their execution, their location and their 'associations'; but the way everything was combined into an experience has gone for ever. Without informants, we can use the art only as a source of hypotheses, never as a confirmation of them; without the artists, our chances of correctly interpreting the content and meaning of a decorated cave are very slight – and how would we know if we were right or not?

Nevertheless, we are constantly learning more about the art and the people who produced it; every new discovery, every new idea add to the picture we are building up. Despite its frustrations, the study of Palaeolithic art is extremely rewarding, not only through the beauty of the figures and the exhilarating experience of visiting the caves – the very places where the artists worked – but also because it still represents our most direct contact with the beliefs and preoccupations of our ancestors, and therefore constitutes one of the most fascinating episodes of prehistory.

Fig. 119 *Jean Vertut in the Salon Noir, Niaux (Ariège). (JV)*

Appendix

by Alexander Marshack

Jean Vertut had intended to write an appendix to this book, giving an account of the methods used in his work in the caves. As he is unable to do so, Alex Marshack has stepped into the breach with the following text concerning Jean's methodology. He presents a transcript of a discussion recorded in the USA, and published here by kind permission of David Abrams.

I

For almost a dozen years, Jean Vertut, one of the leading engineers and theoreticians in the field of robotics, would fly to engineering and nuclear laboratories in the USA to discuss developments in the subject and research projects being conducted between France and the United States. On each such visit, he would stop for some days at our apartment in New York, arriving with a valise of documents, a loaf of French farm bread, fresh camembert cheese, a sealed packet of strong French coffee, and hundreds of transparencies documenting his recent photographic work in the Upper Palaeolithic caves. In his last years this included his work in the Volp Caverns, Trois Frères and the Tuc d'Audoubert. The engineering and robotics projects represented his theoretical and practical professional work, part of the sophisticated contemporary effort to replace man in certain specialised areas of production and mechanical motion. The study of Upper Palaeolithic art represented Jean's equally professional and technical, but private, personal involvement in the effort to understand the work of the hand and mind of early man, particularly as found 'at home' in the Franco-Cantabrian region of France and Spain.

On these visits we recognised that these areas of Jean's work and interest both equally represented the capacity, mind and work of man. We would talk without cease for hours about the theoretical and practical problems involved in robotics and engineering. These discussions included the extraordinary complexity of dealing with the real physical world with instruments and sensors, and the even more complex problems of dealing with the real physical and cultural world of man through concepts and symbols. For Jean and me, these problems, if not equivalent, were comparable. We went imperceptibly from one field of discourse to the other. Both theory and technology formed the base of our discussions within these areas. The constraints and limitations present in these human efforts were discussed, as well as the techniques and potentials for increasing analysis, understanding and production. It is doubtful whether the Upper Palaeolithic symbol systems and cultures were ever discussed in quite this philosophical and technological manner or within such a context.

Jean brought to these highly technical and philosophical discussions a deep interest in religion, that is, in the affective, emotional contents of symbolic imagery and ritual. He once brought with pride a recording in which he, his wife and children, as well as others, had participated in the reconstruction and production of early Christian songs and hymns. In our discussions about image and symbol, Jean referred often to the work of Marcel Jousse, the French Catholic philosopher and semanticist, whose earlier theories on imagery, symbolism, information exchange and education, Jean suggested, had apparently influenced

André Leroi-Gourhan, whether consciously or not. We had long discussions about the levels of information in image and symbol, and how these changed during production, during repeated viewing, and during analysis. I added to these discussions an interest in neuropsychology and cognition, in the way in which the human hand and mind work to solve problems in the real and cultural worlds, and the many levels of information that are contained in any cultural set of images and symbols.

In the evening, when the sky and room had darkened, we would project our hundreds of transparencies and photographs, discussing the data or information apparent and missing in each photograph and image. We discussed the techniques that might be used to secure different levels of information from the same image. In the 1970s I had pioneered the use of special analytic techniques for study of the cave images, as I had earlier pioneered the use of the microscope to study the problems of manufacture and use among the mobiliary symbolic materials. With grants from the National Science Foundation and the National Geographic, I had developed the use of infra-red, ultraviolet, and fluorescence techniques, as well as the technique of microscopic photography for the study of a host of specialised problems in the caves. Each image, panel and cave represented different sets of problems and required different sets of analytic techniques. I had deposited a set of results with the Minister for Cultural Affairs in Paris, but also lent Jean my filters, lamps, charts and results, and he tested these in his own cave work. We compared results in these evening discussions, suggesting to each other still newer modes of analysis and documentation.

Jean had pioneered in the effort to recreate the ambience or atmosphere of the cave environment, the overall impression of three-dimensional space and complexity that could not be captured in the flat photographs of animals or details of tableaux. His photographic reconstructions of the ceiling of bulls in Altamira and the panel of horses and cattle in Lascaux, prepared for Leroi-Gourhan's classic volume *Préhistoire de l'Art Occidental*, remain the most widely known images of these two famous tableaux from the two giants of Franco-Cantabrian cave-art. In the 1960s he photographed the famous clay bison in the Tuc d'Audoubert, using two lamps. Later he photographed the bison using four lamps, a dramatic three-dimensional photograph that has become the classic representation of these famous clay-carvings. The Lascaux panel and the Tuc d'Audoubert photographs were key images in the first exhibition of European Upper Palaeolithic art in the United States, which I prepared, with Jean's help, in 1978. In the 1980s, while

working with Robert Bégouën to document the caves of Tuc d'Audoubert and Trois Frères, Jean photographed the clay bison in a still more complex manner. He used six lamps to enhance the three-dimensional rendering, photographed the bison in stereoscopic photography, made a wide-angle documentary photograph of the clay bison in their context within the low-ceilinged chamber, for the first time giving us a sense of their size within the chamber (they had seemed over-large and life-sized in the earlier photographs), and finally he documented them in photogrammetry, and then, by computer use of the data, helped reconstruct a model of the bison for exhibition at the Musée de l'Homme in Paris.

And still Jean was dissatisfied. He made endless close-up, detailed photographs of the bison, documenting the way in which the statues were carved, using tools and hand-modelling, documenting the remnants of unfinished clay left on and around the carvings, the evidence of bat clawmarks around the bison, etc. In this sense, no single photograph was a final or adequate document. Jean and I had for years insisted that the usual flat photographic document of the cave image and the even more schematic tracing or 'relevé' were inadequate for documenting the analytic complexity to be found in the images and tableaux. It was, therefore, with a sense of profound appreciation that we welcomed and discussed the analytical work of Brigitte and Gilles Delluc in deepening methods of analysis and documentation in their studies of limestone carvings in the homesites and caves of the Aurignacian period in the Dordogne. We welcomed this work as another step in the direction of the necessary methodological analysis that went beyond the simple visual surface observations and analyses of the earlier researchers.

Jean had done the photographs for Leroi-Gourhan in the 1960s and had slowly come to realise that these photographs, while dramatic and beautiful, were also 'falsifications' and subjective 'abstractions', and that one must, in the next stage, go beyond these simple surface images or 'documents'. He attempted to deepen photography as a documentary means, and therefore welcomed the cartographic and illustrative method, developed by the Dellucs, of providing information that supplemented the photograph. I had experimented with comparable but simpler techniques of schematically rendering the analytical data that is documented in the photograph but is not apparent to the untrained observer. For these reasons, Jean had arranged a meeting with the Dellucs in Périgueux in 1983, where we discussed technology, theoretical, analytical and interpretative problems; and a need for a joint effort by researchers in the field to discuss the problems and possibilities being opened up in the contemporary period.

Leroi-Gourhan had instituted an earlier stage in the investigation of the cave-art by his visual and structural studies of the images; the relation of the images to each other; and their position, form and style. While Jean had provided most of the photographic documentation for Leroi-Gourhan's volume, Leroi-Gourhan's work was not analytical in the sense of the work that Jean, among others, then attempted to develop; no single image or composition was given intensive internal study. The young researchers who followed Leroi-Gourhan (including his students the Dellucs, Michel Lorblanchet and Denis Vialou, and an older generation including Jean and myself) had begun experimenting with new levels of analysis and interpretation. Jean, as an advocate and discussant, was attempting to circulate information on these many efforts among the researchers and to bring together the separate persons and technologies. Many of the symposia and colloquia that began to appear were in some measure instituted by his persistent advocacy and efforts.

There was no better opportunity to appreciate Jean's perception of the caves as an area of analytic problems than a visit to one with him. I remember a decade ago in one deep chamber in the Tuc d'Audoubert that we came upon an unpublished 'Aurignacian' panel of intertwined incised *macaronis* or meanders. He and I lay down on the damp clay floor and, like students at work on a puzzle, went line-by-line through the entire composition, attempting to determine the point of beginning (apparently within a crack in the wall) and to reconstruct the sequence and logic or 'strategy' of its subsequent development or construction. The *macaroni*, exiting at the left from the crack in the wall, then wandered, twisting in every direction, intertwined and interlaced, and crossed over to the other side. We lay on our backs for a long time, discussing how to unscramble and document the composition, and what photographs would be required at points of intersecting and overcrossing lines to determine priority in engraving. In traditional documentation by tracing or photography, such a panel would have been published as a meaningless mélange of wandering lines (see the abbé Glory's tracings of the Magdalenian *macaronis* in the volume *Lascaux Inconnu*, edited by A. Leroi-Gourhan and J. Allain, 1979).

Jean later began the photographic documentation and we discussed his first photographs and the need for closer analysis, but he died before its completion. The *macaronis* in the Tuc d'Audoubert were perhaps the least interesting evidence of human activity in the cave in the traditional frame of archaeological evidence but, like sets of intentional fingerprints and spear marks found in the clay in other parts of the cave, they added to the sense of variable and complex symbolic use to create at times images of 'art' and, at other times, images of participatory ritual marking.

In the same cave, we also squeezed into a tiny, difficult-to-reach 'chamber' whose upper portion was large enough only for our two heads and our arms which were used to point out the images and the problems we would like to document by photography and microphotography. The chamber, with its rounded ceiling, had at its centre a tiny, crudely engraved horse encircled by a large number of 'P' signs, referred to as a feminine sign in Leroi-Gourhan's classification, a schematic figure apparently derived from the female form. Jean and I examined each sign, trying to recreate its mode of manufacture. Some were clearly made by stone tools, but others by a finger; some were over-engraved or renewed; most were made by right-handed persons, some by the left hand; most were in a sequence or row of 'P' signs, though some, apparently added later, were isolated. We stood there and discussed which images needed microscopic documentation, how the entire composition and sequence of manufacture could be reconstructed, and what levels of information required documentation.

Jean began the documentation we had discussed, and the following year we spent one long evening going over his preliminary photographs and sketches of the composition. This was not the Jean Vertut of the beautiful photographs but Jean the scientist, analyst and investigator, the technician and theoretician. The excitement of these developing analytic efforts and his increasing steps to document the dynamics of cave use and symbolism made him speak often of retiring early from engineering to take a degree in Palaeolithic archaeology, and devote all his time to the study of the art. He died in mid-career, both in the field of robotics and in the study and documentation of the caves.

II

Jean Vertut had begun taking photographs of the art in the caves in mid-century, before the introduction of modern flash and strobe or of high-quality colour films. With each introduction of technology, he drew on his knowledge of engineering to update his cameras, meters, lights, films, and eventually began to use special filters and films. A large part of our discussions in the last few years concerned his use of new techniques and equipment for the analysis or documentation of particular types of problems. At times he developed his own photographic apparatus and technical devices when these were not commercially available. He had begun to use infra-red and ultraviolet in conjunction with special filters, in part following on and enlarging on my earlier efforts. He also began using high contrast colour films, developed for biological microscopy, to enhance the cave images. He did macroscopic studies of the engravings to study faint and difficult images. He experimented with stereoscopic photography and three-dimensional representations of carved and sculptured forms. He conducted photogrammetric studies of the bison of the Tuc d'Audoubert to create a computerised reconstruction of the sculptures. He developed a method of piecing together major cave tableaux (Lascaux, Cougnac, etc) from separate photographs by using arrays of projectors to create overlapping seamless images. He had begun to use Munsell colour charts in recent years to verify the accuracy of the colours in his reconstructed panels and cave tableaux, but he also intentionally altered the colour values of certain images to increase contrast or heighten faded images. We had both begun an investigation of the uses of computer enhancement to recover images and to unscramble particularly complex accumulations, and were preparing a program to initiate use of the method for specific analytical problems in some of the major caves.

Jean never developed techniques merely for their use but always as a means of attacking and solving particular problems. He not only pioneered in the use of new techniques but encouraged their testing by colleagues. There was never a sense of proprietorship or secrecy in his development of new approaches. This combination of technical advocacy and an ongoing effort to document and present the cave images at different levels of 'information', as he put it, made him one of the pioneers of the present new stage of research and documentation, a research which he had begun when assisting Leroi-Gourhan in mid-century and continued to develop until the time of his death. Without offering a hypothesis as to the meaning of Upper Palaeolithic cave-art, he contributed enormously to the documentation and to our understanding and appreciation of its complexity and beauty.

III

Some months before he died in 1985, a videotape was made in California of Jean's informal comments and thoughts on Upper Palaeolithic art and his developing analytical studies. The tape was made at the home of David Abrams, an American who for a number of years has led some of the American tourists who pass through the Franco-Cantabrian caves. Jean's responses to Abrams's questions, while informal and sketchily conversational, reveal the wide range and depth of his

practical and theoretical concerns, both in documenting and in attempting to understand the dynamics of the caves and their uses. His answers to Abrams almost always attempted to indicate the inadequacy of traditional modes of documentation and inquiry, and they reveal his own developing interest in going beyond the flat, if magnificent, photographs of single images for which he had become known, particularly as the major photographer of Leroi-Gourhan's volume. The videotape is a record of Jean's many levels of interest in the caves, and the diverse approaches he was using to document them.

The tape has been edited somewhat because the informal mealtime conversation during which it was made, and Jean's own occasional search for an expression in English to convey an idea that had come to him in French, tended to break the flow of his thoughts. I have here and there added the unexpressed information needed to clarify a thought.

To a question from David Abrams: 'The question reminds me of Marcel Jousse ... Leroi-Gourhan, incidentally, denied that he had read Marcel Jousse. It may be that he forgot. Marcel Jousse was a Jesuit, a priest, who had attempted to take a direction taken by anthropologists today [that is, a study of the way in which information is both remembered and exchanged – A.M.]. Jousse learned the ancient Latin and, by going to the early Biblical texts, he tried to indicate the manner in which Christ, for instance, taught his followers. Jousse attempted to "play back" Christ's words as they are presented in the Bible (within their original cultural context) and to give them a meaning greater than they at first appear to have. In 1965 Jousse presented what may have been the first anthropological theory of communication. He wrote four volumes called *L'Anthropologie du Geste*, the "Anthropology of Gesture" ['gesture', however, not in its English meaning but in the French sense of providing meanings beyond the overt significance of the words – A.M.]. Jousse used the term "anthropology" in the way it is used in English, not in the way in which it is used in French. He was interested in the fact that information and memory proceed by way of continuous "playback" – by the "playback" of emotion, by the "playback" of the word, by the "playback" of the image, and so on, and that it is by such continuous and repeated playback that we reach a higher level of knowledge [that is, a synthesis of meanings at a higher level – A.M.]. This occurs not by a straightforward intellectual approach but by an incremental method which is, incidentally, the way that one actually learns ... One learns by way of the image, the gesture, and the word. There are many levels of information in these modes ... and also, therefore, many levels of analysis. In the study of Upper

Figs. 120/121/122 *Different views of the clay bison of the Tuc d'Audoubert (Ariège), showing how they were made and set upright against the rocks. (JV). [See also Figs. 61 and 62]*

Palaeolithic art it was Alex who put me on to this track ... The Dellucs have recently added to this type of analysis of the images, and Robert Bégouën has also been a powerful influence in developing new levels of analysis. There are many prehistorians of cave art who still think that the "*relevé*" (the tracing) is the real (or valid) level of information for the cave images.'

QUESTIONER: 'You talked about the importance of "gesture" in symbolism. Do you mean by "gesture" the actual act of painting?'

VERTUT: 'This is the first level only of what I mean. There are, necessarily, other levels. I am not sure that "gesture" in English has the same meaning as "*geste*" in French. "*Geste*" in French has a greater power. It has at the same time the meaning of emotion [effect and emphasis – A.M.]. It may refer to the roots of strong emotion that were always present in the ancient oral tradition. When you hear the mere recording of a text, for instance, you lose a significant part of the original meaning. Such a recording, of course, is better than losing the whole ... The fact that one draws on a wall does not provide the same meanings as would the gestures that would accompany your words or the meaning of the words themselves. These are difficult and profound questions ... They concern the differences in teaching by a use of words, images, or gestures and emotions ...

'After the discovery of Lascaux, Marcel Jousse attempted to go beyond Breuil's limited approach to the images by giving additional life and depth to their possible meaning, stating that they were "mimograms" [images containing complex levels of information – A.M.]. Of course they were mimograms. But of what type? That question could not be answered in the Forties because archaeologists at that period believed that a particular image, for instance, was a "shaman" or a "hut", applying ethnographic analogy and ethnographic information to the images in the caves. The problem is far more complicated...

'With Leroi-Gourhan's work it was finally realised that Upper Palaeolithic imagery was, in fact, more complicated. Actually, it is as complicated as religion itself and the most complicated religious imagery and symbolism. It represents, in fact, another universe. We do not today live in that universe ... Leroi-Gourhan was the last in the generation of what we call in French "*littéraires*", who thought of Upper Palaeolithic art as the beginning or introduction of art, as part of the history of art. You know, of course, that the history of art is a "game" that provides us only with certain types of analytic information, but it is not the basic or original information contained in the image itself. My wife, for instance, has been painting icons. The way in which the historians of art attempt to explain these icons makes no sense to someone who knows their origins and contents. Religion helps ten times more in giving them meaning.'

ABRAMS: 'You don't know how far one has to go to find anyone in the United States who thinks this way, much less talks this way about cave-art.'

VERTUT: 'Alex does this ... I think that you, too, are probably moving in this direction ... in publishing and in trying not to criticise but to advise most archaeologists who are interested in cave-art, please do not go to the drawings of Breuil. The printed [archaeological] material often contains too little information to discuss this art intelligently. One must go to the material and do better research ... everything that has been published by researchers in cave-art ... may be, in some measure, true, but the truth is, in fact, far more complex.'

ABRAMS: 'I have some pictures here of Altamira made by Breuil. I now wonder how much the pictures can be trusted ... They are so ...'

VERTUT: 'You have to understand that he was interested in one image at a time. His effort was to remove the one image from its context, separating the images [documenting each image – A.M.]. He was, at that time, thinking of the "art for art's sake" explanation, which was part of the problem. It took Breuil a long time to see and document everything separately. If you use these separate images merely for guidance, and to keep a record and not as the fundamental information but as guides to memory which can be useful in reconstruction ...'

ABRAMS: 'Jean Noël, a guide at Niaux, took us to the Salon Noir. He didn't say anything. He started to sweep with his lamp the whole side of the cave ... He brought me to the concept of not the individual painting but the panel ... It was the most dramatic ...'

VERTUT: 'I had a marvellous visit last summer in Niaux with young friends I met in Niaux with Clastres when he was alive. I always work in Niaux in the night time. I spent many nights, sometimes six or seven nights, in 1961 and 1962 and '64, and then later again in '67. I had the privilege to visit the Clastres Gallery with Leroi-Gourhan and to photograph the hall with footprints in the sand ... That was very impressive ... I have' documented about 95% and reconstructed some of the panels ... I did it again in 1972 for Le Thot [the Centre for Prehistoric Art, in the Dordogne].'

ABRAMS: 'The bison in Niaux in the book. That picture is extremely powerful. It has a power as big as the book itself …'

VERTUT: 'I had a Graflex at the time … This is the simplest technique of just taking the picture. The subject is powerful. The picture is just a good picture. This is the easiest gesture to make. In these days there is no difficulty at all in taking a good picture … We have high resolution and great enlargement … That profile of the bison is about full scale. There is great difficulty in photography. When you take a picture, you keep certain things about the "*graphisme*" [or image itself] but you lose others. I did for Robert Bégouën the first microphotographic enlargements, three to four times natural size, of engravings [in Trois Frères and the Tuc d'Audoubert]. He was amazed at what could be seen … If you want to reproduce the Lascaux paintings, for instance, natural size is sufficient, but if you want to do a replica of the surface, it must be done in the small tunnel where you descend to the red deer … I think you must produce a 3× target replica … but this again is merely a technique.'

ABRAMS: 'I think that some of the power of the Niaux bison is their full size…'

VERTUT: 'But that is only *one* of the factors …

[Later] 'This is one of the particular ways of drawing the bison which is not an exact bison but is instead very human-like. The eyes are human eyes … You have two types of bison in cave-art. There are those that have nearly human features … In Altamira in the deep gallery you have reliefs where the added eyes and horns … [make it] impossible to tell whether the head is a human or bison … but here in the Breuil drawing of an Altamira bison the individual is complete … This is one of those interesting ambiguities. When everything is clear and simple, it is not interesting … And this is one of the interesting things about the cave-art … The "woman-bison" of Pech-Merle just underneath the red hand and dots, as seen by Leroi-Gourhan, and the woman on the ceiling made by finger-marks, apparently represent a form of transformation … But this is full of ambiguity … You know, Leroi-Gourhan found the curious relationship of female figures and bison, but we still do not know if the horse is related to the male … He found many examples where the female and bison were associated, and this was one reason he suggested that the bison was "sexual" and female. But the horse [which Leroi-Gourhan suggested was male] is ambiguous. The female and bison association is frequent but not the male with the horse … There are a few examples: in Lascaux, Villars and Trois Frères …

This is exactly the same tradition of "*graphisme*" that Alex is studying.

'Concerning the head of the Niaux bison which opens Leroi-Gourhan's volume and which you [Abrams] find so powerful, this is precisely the point at which trouble starts. To me your reaction to the head of the bison is not excessive, but the full bison is far more interesting than the face … Showing the face at full scale [as it appears in the book] favours the emotional reaction to it. This is, of course, accepted, but if you read too much into the emotion you can disadvantage your approach …

'So to substitute a cave visit is one of my aims … to show the art properly to the public. The only way to do this is to show a full-scale print … This was attempted approximately at Le Thot, but it is also necessary to properly prepare a person before he sees the picture. That is, of course, impossible … In fact, it is only during recent years in the preparation for Le Thot that I began to try to counteract the problem of growing difficulty in visiting the caves.'

ABRAMS: 'Some years ago there were no pictures of people in the caves to give a sense of scale to the images.'

VERTUT: 'This is a problem of the photography …

[Later] 'Back to the scale problem. You don't need people in the picture for scaling as far as I'm concerned. Those who have once been in a cave already have a sense of scale … We have to reveal the scale carefully in another way … Those who have never been in a cave on seeing the pictures lose their sense of scale and … time [the spatial/temporal sense of distance, size, depth, etc. – A.M.]. Those who live in a structured environment [have their own normal sense of time and space from their environment].

[Showing a photograph taken in the Tuc d'Audoubert] 'These are finger marks [in the clay floor] … We can see them in detail now, made with my zoom so the field is very limited … It's very difficult to see with flash … This is the closest I could go without a tripod. I never use a tripod because it is too time-consuming … This was thought to be a bear footprint … When you get close up, you can see the traces of the bear hair in the clay and there are hundreds of them in the cave … Here is a bear skull on the ground. Next to it a [Palaeolithic] person came, and here are the traces of knees on the ground. The person broke off the bear canine which is the "nice" tooth. This is the act of a few seconds which you have to trace [and document by photography – A.M.]. Just two metres away is a small "*gour*", a small dam [or clay mound or cone]. Here at the right you see the slippage of the right footprint … other footprints here … and you have the skeleton of a snake that was left. One of the vertebrae was recently removed and studied, and it was found to be a viper. You have many traces of this kind which are traces of life … Further on you have footprints, heelprints … This is a rib that was removed by prehistoric man from a very old skeleton of a cave bear … A little further on … footprint slippage … with holes from water dripping from the ceiling.

'The visit to this cave, and one or two others, was to a very secret place … where maybe once three or four people came and left … Closer to the clay bison is a deep depression in a small chamber called the "heel chamber". The clay has been covered by calcite, some of the calcite has been removed [in modern times] and underneath you see absolutely fresh clay still soft where finger marks are still fresh. The problem is how to study this area of human activity without destroying all of the covering…

'Here are heelprints in a very low-ceilinged room. Here again you see fingerprints, as typical in the Tuc [d'Audoubert], double prints of the fingers … Here is a sign like the branch of a tree that Alex would call "vegetal" … The full study of the area has still to be done …

'This is a big, big vulva that is in the book … I am now preparing microphotographs by which it will be possible to identify the types of tool used … On the right is a small mound, apparently made by humans since it is impossible otherwise to account for the mound in an area where everything else is flat. On this mound there are [spear] marks. One can't walk in this area because it is covered by calcite so the pictures were taken at some distance.

[Referring to another photograph] 'In this picture is a stalagmite that was used to cut the clay and remove a big plate … approximately in the shape of a bison … Here the finger marks again, so this was attended with a lot of ceremony. On the other side there are pieces of *macaronis* which were covered with clay. These were tests in my campaign to indicate the way in which we could work … to get more information or data on the footprints in a precisely measured, good, metered way for photogrammetry construction. Here are the spear marks and scratches in the mound. It will, however, be possible to make a mould and get a print of the spear.

'The original photo of the two bison was made with two flashes … but the magnificent picture you saw earlier was made with four flashes. Here is a picture of the rock upon which the bison were placed. The two bison were, at first. badly supported, so support clay was added below … Here is a photo of the female bison. We can see the horns and ears, and the photo reveals the complete process of finishing it. These marks were made with tools, but these others by fingers (Fig.62).

'This is a long-distance shot of the chamber with the two bison. It was made with a wide-angle lens. This is a unique picture never made before. It gives you a detailed view of the ceiling including much new information. You can see the red of the ceiling which may have been one of the reasons for placing the bison here (Fig.61.)

'Here is the picture of another bison, a very small third one, behind the other two. This bison is a duplicate made from clay of the cave, a duplicate that has been officially returned to the cave in a casting made at St Germain [from the original] by a special process.

'Here is a close-up photo of a piece of clay that had been left on the rock with all the fingerprints still on it. The carvers left it there after the bison were finished.

'Here is a picture of the supporting clay under the bison. Here is a close-up of the broken tail. Here on the rock and the bison are small, intriguing marks. They seem to be marks made by bats that landed on the sharp edges.

'Here is the photo of an extra piece of clay that was left at the spot [by the original makers] that should have been removed at that time.

'Here is the photo of a very dramatic recent discovery. On the supporting rock opposite the bison we found the only engraving. That engraving, a "sign", is like a signature.'

Almost every picture that Jean took in the caves of the Tuc d'Audoubert and Trois Frères in the last few years, while working with Robert Bégouën, was taken to make a point, or to present data on the complexity of the human activity associated with use of the caves and the creation of the diverse images and symbol systems found in these major caves. When he showed his slides, there was a lesson in documentation, analysis and complexity in each. There is a profound sense of loss in his passing and the silence that now accompanies his massive documentation.

Notes

Preface

1. See discussions and references in Davis 1986, 1987;
   Bednarik 1986; Pfeiffer 1982; Stoliar 1977/8, 1981;
   Ucko 1987; Zhurov 1977/8, 1981/2.
2. Sieveking 1979.
3. Pfeiffer 1982.

Introduction: Cave Life

1. Andrieux 1974, and pers. comm. See also Pales & de
   St Péreuse 1976, pp. 129, 133; 1979, p. 140; Pales *et
   al* 1976.
2. *L'Art des Cavernes* 1984, pp. 433–7; Vialou 1986.
3. Bahn 1987b; Tyldesley & Bahn 1983.
4. Arl. Leroi-Gourhan 1983.
5. Vallois 1961; Dastugue & de Lumley 1976, pp.
   616/17.
6. Roper 1969, p. 448.

1. The Discovery of Ice Age Art

1. Pittard 1929.
2. Faure 1978, p. 75.
3. Contrary to most textbook accounts, the Chaffaud
   bone was not found in 1834 (see de Saint Mathurin
   1971; Worsaae 1869).
4. Lartet 1861. For his later finds, see Lartet &
   Christy 1864, 1875.
5. Bouvier 1977, pp. 54-7.
6. Kühn 1971, p. 14.
7. Simonnet 1980, p. 13.
8. Contrary to the often repeated claims (eg Nougier
   & Robert 1957) that the 'paintings' of Rouffignac
   were first mentioned by François de Belleforest in a
   publication of 1575, and occasionally by later

   authors, there is not a single mention by anyone of
   drawings in that cave before the mid-twentieth
   century (see de Saint Mathurin 1958; Delluc &
   Delluc 1981).
9. Molard 1908.
10. Chabredier 1975, p. 12; Ollier de Marichard 1973,
    p. 28.
11. Cartailhac 1908, p. 522; Ollier de Marichard 1973,
    p. 28.
12. *Découvertes* 1984.
13. Kühn 1955, pp. 45/6.
14. Maria gave this account in conversation with
    Herbert Kühn in 1923 (see note 13). She claimed
    she was five years old at the time of the discovery;
    other informed sources place her age at eight
    (García Guinea 1979, p. VIII) or nine (ibid. p. 19).
15. Contrary to popular accounts, Maria would not
    have used the word '*toros*' for the animal figures
    she saw (see García Guinea 1979, pp. VIII and 20).
16. Kühn 1955, p. 46.
17. de Sautuola 1880.
18. Kühn 1955, p. 48.
19. García Guinea 1979, pp. 42-4. On de Mortillet's
    anticlericalism, see Reinach 1899 and Bahn 1988.
20. Harlé 1881.
21. The earliest known find of a Palaeolithic lamp
    occurred in 1854 at La Chaire à Calvin (Charente),
    and several others were also discovered before
    Altamira, but they were either contested or
    ignored, while some remained unpublished for
    years because of the controversy over the existence
    of a Palaeolithic lighting system (de Beaune 1987,
    p. 12).
22. Bégouën 1947, pp. 494/5.

23. Kühn 1955, p. 49.
24. García Guinea 1979, p. 42; Madariaga de la Campa 1981, p. 302.
25. Breuil 1964, p. 11.
26. Piette 1889, p. 206.
27. Roussot 1965; Delluc & Delluc 1973, p. 205.
28. As mentioned in note 21, early finds of lamps had met with resistance or ridicule; Rivière himself had originally failed to take seriously the claims of a M. Detrieux who informed him in 1896 that he had found a lamp near Le Moustier (de Beaune 1987, p. 12). In short, the history of the discovery and acceptance of Palaeolithic lamps closely mirrors that of cave-art itself.
29. Roussot 1972/3.
30. Bégouën 1932, p. 8; 1952, p. 289.
31. Cartailhac & Breuil 1905.
32. Cartailhac 1902.
33. Brodrick 1963, p. 63.
34. Cartailhac 1908, p. 520.
35. Cartailhac 1908, p. 512.

2. A Worldwide Phenomenon

1. Bahn 1987.
2. Aveleyra 1965; Messmacher 1981, p. 94.
3. Messmacher 1981, p. 84.
4. Kraft & Thomas 1976.
5. Anati (1986, pp. 789/796) claims that some rock art at Rio Pinturas, Chubut (southern Patagonia, Argentina) is 'associated' with archaeological levels dating to 12,000 years ago, but this remains unsubstantiated. The Peruvian cave of Toquepala has red figures of camelids, deer and armed hunters on the wall, and two small 'brushes' of wood with tips of wool impregnated with red ochre were found stratified here in levels dated to almost 10,000 years ago (Julien & Lavallée 1987, pp. 49-50). See also Linares 1988.
6. Guidon & Delibrias 1986. Bednarik (pers. comm. 27/8/87), who recently visited the site, does not accept this date for the paintings; some of the art is certainly about 9,500 years old, and there is also 'much older art' which remains undated.
7. Whitney & Dorn 1987.
8. On Zimbabwe, see Walker 1987, p. 142. Mori (eg 1974), like Frobenius before him, has proposed that the earliest rock-engravings of the Sahara may date back to the late Pleistocene. However, other scholars such as Muzzolini (1986, pp. 312-14) point out that this reasoning is based on a few isolated, and possibly anomalous, radiocarbon dates from occupations at the foot of decorated rocks, and which may have no association at all with the engravings. The question therefore remains open.

Anati (1985, p. 54; 1986, pp. 789/791) believes that some rock-art in Tanzania may be 40,000 years old, partly because levels of that date have been excavated in decorated sites, but also because some parietal figures seem to resemble those on the Apollo 11 Cave's stones.
9. Wendt 1974, 1976.
10. For Border Cave, see Butzer et al 1979, p. 1212, note 44; for Matupi Cave, see Van Noten 1977, p. 38.
11. Anati 1986, pp. 789/792.
12. Kumar et al 1988; Wakankar 1984, 1985; Anati 1986, pp. 789/793.
13. You Yuzhu 1984.
14. Sohn Pow-Key 1981.
15. Sohn Pow-Key 1974, p. 11.
16. Aikens & Higuchi 1982, p. 107.
17. Wright 1971.
18. Rosenfeld et al 1981.
19. Brown 1987, p. 59; Brewster 1986.
20. Rhys Jones, pers. comm.
21. Chaloupka 1984; Murray & Chaloupka 1983/4. For arguments against, and a reply by Chaloupka & Murray, see Archaeology in Oceania 21, 1986, 140-47. On the problems of identifying species in Australian rock-art, see Chapter 6, p.115).
22. Dortch 1979, 1984.
23. McNickle 1985; Nobbs & Dorn 1988.
24. Hallam 1971.
25. Aslin & Bednarik 1984a/b, 1985; Bednarik 1986.
26. For regional studies, see Barandiarán 1973, Moure Romanillo 1985 (N Spain); Chollot 1964, 1980, de Saint Périer 1965, Piette 1907 (France); Lejeune 1987, Twiesselmann 1951 (Belgium); Adam & Kurz 1980, Bosinski 1982 (Germany and Switzerland); Eppel 1972 (Austria); Graziosi 1973 (Italy); Freund 1957, Hennig 1960, Valoch 1970, Müller-Beck & Albrecht 1987 (Central Europe); Abramova 1967, Praslov 1985 (USSR).
27. Belfer-Cohen & Bar-Yosef 1981, p. 35. There is also a faint possibility that some linear rock-engravings in a number of caves at Mount Carmel are Palaeolithic (see Ronen & Barton 1981); and an enigmatic engraved figurine was recently found in the Acheulian (Lower Palaeolithic) site of Berekhat Ram in the Golan Heights (Goren-Inbar 1986).
28. Campbell 1977, vol. 2, Figs. 102, 105, 143. A brief account, without illustrations, can be found in Sieveking 1972. There is some doubt about the authenticity of a horse-head engraved on a fossilised bone from Sherborne (see Antiquity 53, 1979, 211-16; Nature 283, 1980, 719-20; Antiquity 55, 1981, 44-6 and 219-20).
29. See, for example, Bahn & Cole 1986.
30. Chollot-Varagnac 1980, p. 457.

31. de Saint-Périer 1965.
32. Bouet *et al* 1986/7; Delporte *et al* 1986; a computerised data bank has also been set up for the French parietal sites (Djindjian & Pinçon 1986).
33. Pericot 1942; Llongueras 1972; Fortea 1978.
34. Pales 1969; Pales & de St Péreuse 1965.
35. Bosinski & Fischer 1974, 1980.
36. Bégouën & Clottes 1987; Bahn 1983.
37. Graindor & Martin 1972; Martin 1973.
38. Breuil thought some red streaks in the Welsh coastal cave of Bacon Hole might be Palaeolithic (1952, p. 25), but these have now faded away, and are thought to have been either a natural phenomenon or the traces of someone having recently cleaned a paintbrush on the wall! The recent claims for art on the walls of a cave in the Wye Valley (see *Illustrated London News* Jan. 1981, 31-4; *Current Anthropology* 22(5), Oct. 1981, p. 501-2) proved to be based on a mixture of natural features and wishful thinking (see *Antiquity* July 1981, pp. 81-2, 123-5; *I.L.N.* May 1981 p. 24; *Current Anthropology* 23(5), Oct. 1982, p. 567-9).
39. Maringer & Bandi 1953, p. 23.
40. Bosinski 1982, p. 6; Breuil 1952, p. 24; Freund 1957, p. 55.
41. Bader 1965, p. 25; Maringer & Bandi 1953, p. 27; Anati 1986, p. 789/93; Shimkin 1978, p. 280.
42. Bader 1965, 1967; Shimkin 1978, pp. 280-82.
43. Praslov 1985, p. 181; Abramova, pers. comm.
44. For example, the lone engraved bison of Ségriès (Alpes de Haute Provence) is thought by most (though not all) specialists to be a fake, and was therefore not included in *L'Art des Cavernes*: see de Lumley 1968. For the doubts, see Pales & de St Péreuse 1981, pp. 75-86: the figure is quite different in its proportions, technique and style from unquestionably authentic bison, and is isolated in a region empty of Palaeolithic art.
45. Aujoulat *et al* 1984.
46. *L'Art des Cavernes* 1984.
47. *Arte Rupestre en España* 1987 is the most up-to-date synthesis of the Spanish caves, pending the announced publication of an equivalent of the French Atlas (note 46). See also Straus 1987.
48. Graziosi 1973; Naber *et al* 1976.
49. Basler 1979.
50. Cârciumaru & Bitiri 1983.
51. For detailed regional distribution maps, see notes 46 and 47.
52. Delluc & Delluc 1973; *L'Art des Cavernes* 1984, p. 208/9; Fortea 1981. For other examples of caves with a single figure, see Jordá 1979, p. 332; 1985.
53. *L'Art des Cavernes* 1984, pp. 376/7, 347-9.
54. *L'Art des Cavernes* 1984, pp. 378-80.
55. Leroi-Gourhan & Allain 1979.
56. Vialou 1986, p. 187. It is noteworthy that Vialou (p. 166) counts clusters of the same motif as a single unit: thus the 189 circles engraved on a bear figure at Trois Frères are counted as one sign. If a different system were adopted, the resulting figures and percentages would be totally different.
57. Bégouën & Clottes 1987.
58. Vialou 1986, p. 351.
59. Tosello 1983.
60. Bahn 1982, 1984.
61. Conkey 1980.

3. *Making A Record*

1. Delporte 1979, p. 62.
2. See, for example, Rivenq 1976, which compares modern photographs of a frieze in Montespan with those taken in 1926; Bégouën 1980, p. 689, which does the same for Trois Frères; or Omnès 1982, pl. XIV, which does the same in Labastide.
3. Brodrick 1963, p. 53.
4. Breuil, letter to Glyn Daniel, 20/8/55, in author's possession; see also Aujoulat 1987, p. 22.
5. Bégouën & Clottes 1987.
6. Boyle *et al* 1963, p. 15.
7. Delluc & Delluc 1984, p. 57. Non-existent circles also appear on some of Breuil's versions of portable engravings from Enlène – see Bégouën *et al* 1984/5, pp. 29, 40; this article (pp. 71/2) shows that Bouyssonie's copies of portable art are often more reliable than those of Breuil which feature omissions, additions and a certain amount of wishful thinking.
8. Bégouën 1980, p. 684; Vertut 1980, p. 674.
9. Breuil 1952a, p. 11.
10. Cartailhac & Breuil 1906; Breuil & Obermaier 1935.
11. Breuil 1949.
12. For unquestioning acceptance of Breuil's tracings, see, for example, Leason 1939, p. 51: 'In verifying his theory the Author has had recourse to published sources, but, coming from the distinguished hand ... of Professor Henri Breuil, their accuracy can hardly be called into question.'
13. Pales 1969, pp. 105/6.
14. This was a problem, for example, in a recent re-examination of the heads of the Trois Frères felines – see Bégouën & Clottes 1986/7; and comment by Clottes in Marshack 1985, pp. 175-7.
15. Bégouën 1942, p. 6.
16. Garcia 1979; Martin 1974.
17. Bégouën & Clottes 1987.
18. Pales 1969, p. 40; Airvaux & Pradel 1984, p. 213; Faure 1978.
19. Marshack 1972.

20. d'Errico 1987.
21. Bégouën 1980, p. 685; 1984, p. 79; Vertut 1980.
22. Vertut 1980, p. 667.
23. Marshack 1975.
24. Vialou 1982; 1981, p. 81.
25. Marshack 1975; see also Marshack 1985, pp. 68, 77-9 (and pp. 63/4 for criticism by Lorblanchet).
26. Vertut 1980, p. 672.
27. Lorblanchet 1984; see also Aujoulat 1987.
28. Ruspoli 1987, pp. 177-86.
29. Bégouën 1984, p. 79.
30. Delluc & Delluc 1984a.
31. Laurent 1963, 1971.
32. Tosello 1983, p. 285.
33. Faure 1978, pp. 68/9.
34. Roussot 1984c.
35. Bouvier 1977, pp. 54-7.
36. See Pales 1970 which puts an end to an error of this kind.
37. Lorblanchet 1984.
38. Bégouën 1942, p. 9.
39. Pales & de St Péreuse 1965, p. 221.
40. Pales 1964, p. 12; Pales & de St Péreuse 1965, p. 230.
41. Pales 1969, p. 26; Pales & de St Péreuse 1976, 1981, & forthcoming.
42. Bednarik 1986, p. 35; Meylan 1986, p. 47.
43. Meylan 1986, pp. 46-50; Lhote 1972, p. 323.
44. Ucko 1977, p. 8.
45. Rivenq 1984.
46. Lorblanchet 1981.
47. Vialou 1982.

4. *How Old is the Art?*

1. Airvaux & Pradel 1984; Bégouën & Clottes 1979, p. 19.
2. Piette 1894, p. 129.
3. Bahn 1984, pp. 164/5.
4. Berenguer 1986, p. 669.
5. Faure 1978.
6. Bahn 1982.
7. Combier 1984.
8. The development of Piette's views is explained in Breuil 1909; and see Breuil 1959, p. 15; Delporte 1987.
9. González Echegaray 1968; 1974, pp. 39-42.
10. González Echegaray 1972. A similar attempt to date caves from the dominant species depicted was made by Jordá, quoted in Altuna 1983, p. 236.
11. Bégouën & Clottes 1985, p. 43.
12. Rouzaud 1978, p. 32.
13. Laming-Emperaire 1959, pp. 32/34; Roussot 1984a.
14. Carcauzon 1986, 1988.
15. Gaussen 1964.
16. Straus 1982, p. 78.
17. Combier et al 1958; Combier 1984.
18. Arl. Leroi-Gourhan 1983, and in Leroi-Gourhan & Allain 1979.
19. Cartailhac & Breuil 1906; see also Barandiarán 1973, pp. 302-7.
20. For Labastide, see Omnès 1982, pp. 182, 187; for Gargas, Barrière 1976, pp. 406-9; Ucko 1987, pp. 40-1.
21. de Saint Mathurin 1973, 1975, 1978.
22. Almagro 1976.
23. Utrilla 1979; see also Almagro 1976, p. 99; Straus 1982.
24. Jordá Cerdá 1972, 1980.
25. Cartailhac & Breuil 1905.
26. de Balbín Behrmann & Moure Romanillo 1982; Moure Romanillo 1986.
27. Glory 1964.
28. Leroi-Gourhan & Allain 1979.
29. For example, Breuil 1952, p. 149.
30. Stoliar 1977/8, 1981, 1985; similarly, Hornblower (1951) assigned these models to the Gravettian, but there is no evidence for this either. For a reply, see Zhurov 1977/8, 1981/2; Bahn 1984, p. 343; 1987a.
31. See Altuna 1975, pp. 104/5, or Altuna & Apellániz 1978, p. 22, for an illustration of the difference between a side- and a front view of a living red deer's antlers.
32. Rousseau 1984; Pales & de St Péreuse 1981, pp. 94-7.
33. Guthrie 1984, p. 58; Mithen (1988) has also adopted this view.
34. Riddell 1940, pp. 158/9.
35. Breuil 1977.
36. de Saint Mathurin 1973, 1975.
37. Breuil 1977, pp. 57/8.
38. Ripoll 1964.
39. Breuil 1912; 1952.
40. Breuil & Lantier 1959, p. 225.
41. Laming-Emperaire 1962, 1959.
42. Jordá Cerdá 1964, 1964a; see also Altuna & Apellániz 1976, pp. 158-63; Barandiarán 1973, p. 310.
43. Leroi-Gourhan 1971.
44. Leroi-Gourhan 1962.
45. Delluc & Delluc 1978, p. 386.
46. Moure Romanillo 1985, pp. 124/5; Barandiarán 1973.
47. Delluc & Delluc 1983, pp. 32/3.
48. See Ucko & Rosenfeld 1967, p. 78, for a discussion of these points.
49. See, for example, Bahn 1984 for Pyrenean cultures, or Soffer 1985 for those of the Central Russian Plain.
50. Lorblanchet 1977.

51. Guthrie 1984, p. 72.
52. Almagro *et al* 1972, p. 469.
53. Bégouën & Clottes 1985, p. 49; 1979a, p. 60.
54. Altuna & Apellániz 1976, pp. 157, 165; Apellániz 1982, p. 92; Almagro 1976.
55. Pericot 1942; Jordá 1964; Lorblanchet 1974, pp. 82/3; see also Straus 1982.
56. Airvaux *et al* 1983.

5. Forms and Techniques

1. Marshack 1987.
2. Fages & Mourer-Chauviré 1983; Absolon 1937. It should be noted that the so-called 'Aurignacian flute from the abri Blanchard, located in the Musée des Antiquités Nationales' mentioned by White (1986, p. 109) is actually a specimen from La Roque (Dordogne), probably Gravettian, and housed in the British Museum!
3. Pfeiffer 1982, pp. 181/2.
4. McBurney 1967; Marshack 1987.
5. Allain 1950.
6. Roussot 1970, pp. 9/10. On perforated reindeer phalanges and experiments on their breakage and use, see Harrison 1978.
7. Bibikov 1975. For a cautionary note, see Soffer 1985, pp. 420, 468-9.
8. Dams 1984, 1985, 1987a, pp. 60, 193-5; Glory 1968, pp. 55/6.
9. See Marshack 1981 and Wreschner 1975, 1980 for references.
10. Bordes 1952.
11. See note 9; Mészáros & Vértes 1955.
12. Audouin & Plisson 1982.
13. Péquart 1960/3, pp. 211-14.
14. Bordes 1969; Marshack 1987, 1988.
15. Lévêque & Vandermeersch 1980.
16. Marshack 1987, 1988; Poplin 1983, p. 62.
17. Kozlowski 1982.
18. Hahn 1972; Soffer 1985.
19. Bader 1967 (1976).
20. Praslov 1985.
21. Taborin 1982, pp. 50/51.
22. Poplin 1983.
23. Le Mort 1985; see also Bégouën, Bégouën & Vallois 1937.
24. Taborin 1977.
25. Taborin 1985; Desbrosse *et al* 1976.
26. Soffer 1985, pp. 373, 440; 1985a.
27. Bahn 1977, pp. 252/3.
28. Taborin 1985.
29. Bahn 1982, 1977.
30. Bader 1967 (1976); Soffer (1985, p. 456) estimates only 15 minutes per bead, and hence 2,500 hours for the 3 burials together.

31. Pond 1925; for a more detailed and illustrated account of Aurignacian production of ivory beads, see Otte 1974.
32. Taborin 1982.
33. Jia Lanpo 1980, p. 52; *Atlas of Primitive Man in China* 1980, pp. 114/15.
34. Leroy-Prost 1984, p. 45.
35. Frolov 1981, pp. 63, 77; 1977/9, pp. 153/4, 85. It will be recalled (Chapter 2, Note 27) that an engraved stone figurine has even been found in an Acheulian site in Israel.
36. Marshack 1972, pp. 255-8; Faure 1978, p. 65.
37. Omnès 1982, p. 185; Bosinski 1973, p. 39. Marshack has recently admitted (1985, pp. 72/3) that his own experiments showed that a line's cross-section altered as the line/tool changed direction; it is the changes of direction and of rhythm which suggest to him the presence of different tools or hands.
38. Bosinski & Fischer 1974, pp. 5/6; Faure 1978, p. 65; Marshack 1985, pp. 72/3.
39. d'Errico 1987.
40. Bégouën & Bégouën 1934; L. Bégouën 1939, p. 298; Bégouën *et al* 1984/5. See also de Saint Périer 1930, p. 81; Capitan & Bouyssonie 1924, p. 40; and Pales & de St Péreuse 1976, p. 17.
41. For Ferrassie & abri Durif, see Pales & de St Péreuse 1979, pp. 137/8.
42. Bégouën *et al* 1984, p. 145; many of the La Marche slabs are also engraved on both sides – see Pales & de St Péreuse 1981, p. 22.
43. Omnès 1982, p. 185; Bahn 1983.
44. Péquart 1960/3, p. 222; see also Pales & de St Péreuse 1981, p. 22, for the situation at La Marche.
45. Omnès 1982, p. 184.
46. Bégouën & Clottes 1981, p. 42; 1983; 1985.
47. Moure Romanillo 1985, p. 103.
48. Rozoy 1985.
49. Bosinski 1973; 1984, p. 318.
50. Moure Romanillo 1985, p. 103.
51. Pales & de St Péreuse 1979, p. 141; 1976, p. 17.
52. Couraud 1985, 1982; Bahn & Couraud 1984; Bahn 1984a.
53. Thévenin 1983; Couraud 1985.
54. Rozoy 1985.
55. Bosinski 1973, p. 41. See also Russell 1987, pp. 206-9, for similar experiments.
56. Delporte & Mons 1980.
57. Bégouën & Clottes 1985, p. 45; Pales & de St Péreuse 1979, pp. 141/2.
58. For painted bones, see Gladkih *et al* 1984, p. 142; Soffer 1985, pp. 78, 84; and Shimkin 1978, pp. 275, 283; for paste inside engraved lines, see Marshack 1981; 1979, p. 288.

59. Péquart 1939/40, p. 451; Almagro 1976. For Le Mas d'Azil, see also Pales 1970; Delporte & Mons 1975. For a brief review of the topic, without illustrations, see Sieveking 1983. For plaquettes as a whole, see Sieveking 1987.
60. de Sonneville-Bordes 1986, p. 633.
61. Almagro 1976, p. 68 note.
62. Delporte & Mons 1980, p. 43 – this paper also gives references to some of the earlier experimental work by these authors. For the nineteenth century experiments, see Leguay 1877, 1882.
63. For Pech de l'Azé, see Bordes 1969; Marshack 1977, pp. 289-92; 1976, p. 140; 1976a, pp. 278/9. For Bilzingsleben, see Marshack 1981, p. 189; Müller-Beck & Albrecht 1987, p. 68; and Mania & Vlček 1987, pp. 41, 43; and Mania in *Rock Art Research* 5(2), 1988 (in press). Feustel (in Müller-Beck & Albrecht 1987, p. 60) even claims that marks on one Bilzingsleben bone may be the depiction of a large animal.
64. Leonardi 1983.
65. Marshack 1976, p. 139; 1976a; Freeman & González Echegaray 1983.
66. Marshack 1976, pp. 139-41; 1976a, p. 277.
67. Frolov 1981, pp. 63, 77.
68. Barandiarán 1971.
69. Breuil 1909, p. 380.
70. Bahn 1984, p. 116.
71. Baulois 1980.
72. Allain & Rigaud 1986; Praslov 1985, p. 187.
73. de Sonneville-Bordes 1986, pp. 634-6.
74. Bellier 1981/2, 1984.
75. Barandiarán 1968; Sieveking 1971.
76. Jelínek 1975, pp. 458/9.
77. Passemard & Breuil 1928.
78. St Périer 1930, p. 97; 1936, pp. 116-20; Bellier 1984, p. 26.
79. For the new Spanish specimens, see, for example, Fortea 1981; for Labastide, see Simonnet 1952; Omnès 1982, p. 192.
80. Bellier 1984, p. 30.
81. For the French and Swiss type, see Bandi & Delporte 1984; for the relationship with antler shape, see Leroi-Gourhan 1971, Figs 34/35. Not all compositions of this type, however, were determined by the antler's shape – see Pales & de St Péreuse 1965, pp. 227/8.
82. Cattelain 1977/8.
83. Bahn 1982, pp. 257/8; 1984; Robert *et al* 1953.
84. Péquart 1960/3, p. 299.
85. See Bahn 1978; Bahn & Otte 1985; Péquart 1960/3, p. 219; Vasil'ev 1985.
86. Klíma 1984, 1983, 1982.

87. For Isturitz, see St Périer 1930, pp. 109-16; 1936, pp. 126-33; Mons 1986. For Bédeilhac, see Bahn & Cole 1986.
88. St Périer 1936, pp. 123/5.
89. For a fine example, see Pales & de St Péreuse 1979, pp. 118/19; on the topic as a whole, see Chollot-Varagnac 1980, p. 452.
90. Arambourou 1978, pp. 116-24.
91. Eppel 1972, p. 78.
92. Hahn 1979, 1982.
93. Delporte 1979.
94. de Saint Mathurin 1978.
95. Breuil 1959, pp. 15-17 & 109; Breuil & Lantier 1959, pp. 193/4. In Breuil 1909, p. 399, he states that Piette's 'Vallinfernalien' layer (= Gravettian) was above the '*Eburnéen ancien*' layer containing the ivory sculptures, which must therefore be pre-Gravettian. Yet Delporte (1979, p. 221) claims that there is no evidence of any French statuette being Aurignacian! Other Breuil references to this topic are cited by Pales 1972, p. 250. However, the recent excavations by Delporte at Brassempouy have found ivory fragments only in the Gravettian.
96. Marshack 1984; 1976; 1976a; Bosinski 1982; Hahn 1984.
97. Hahn 1971; Bosinski 1982; Marshack 1987, 1988.
98. Dauvois 1977; for other observations on how female statuettes of ivory and stone were carved, incised and polished, see Delporte 1979, pp. 249-51.
99. Jelínek 1975, p. 411.
100. Praslov 1985; Abramova 1967.
101. Tarassov 1971; Delporte 1979, p. 178.
102. Delporte 1979.
103. Gamble 1982.
104. Soffer 1985; Gladkih *et al* 1984; Jelínek 1975.
105. Bosinski 1984, p. 315; Bosinski & Fischer 1980, pp. 47, 99, 119 & pl. 101.
106. Leroi-Gourhan 1978; for other examples of distortion to fit available space, see Pales 1976/7, pp. 90/91, and Pales & de St Péreuse 1976, p. 53, and 1981, pp. 128-31.
107. Berenguer 1986, p. 665.
108. Marshack 1987, 1988.
109. Delluc & Delluc 1978, p. 393.
110. Delluc & Delluc 1978, pp. 428, 391.
111. Delluc & Delluc 1984b; 1978, pp. 215-21.
112. Lejeune 1981 & in press; Eastham 1979, pp. 367-70; see Vézian 1956 for a variety of examples in one cave.
113. Bednarik 1986, pp. 44/5; see also Breuil 1926, p. 366.

114. Marshack 1977. For Gargas, see Barrière 1976; for Pech Merle, see *L'Art des Cavernes* pp. 470/71.

115. Bednarik 1986, pp. 43/4.

116. Delluc & Delluc 1983a; 1985, p. 60; Bednarik 1986.

117. Vertut 1980.

118. Delluc & Delluc 1983, pp. 58/9.

119. Bégouën & Clottes 1987, 1980.

120. Allain & Rigaud, in Leroi-Gourhan & Allain 1979, pp. 103, 106.

121. Bégouën 1925/6, p. 508; Bouillon, quoted in Ucko & Rosenfeld 1967, p. 241. The idea was first put forward by Max Bégouën, and used in his prehistoric novel *Les Bisons d'Argile* (Fayard: Paris, 1925).

122. Eastham & Eastham 1979, pp. 375/6. On the subject of light sources and visibility of engraving, see also Pales & de St Péreuse 1965, pp. 227/8.

123. Bégouën 1931.

124. Almagro 1981, 1980, 1976.

125. For Fontanet, see *L'Art des Cavernes* 1984, p. 435; for Montespan, see Rivenq 1976, 1984.

126. Breuil 1952a, p. 12; Beltrán *et al* 1973.

127. Beltrán *et al* 1967.

128. Beasley 1986; Bégouën, Clottes & Delporte 1977.

129. Bégouën, Casteret & Capitan 1923.

130. Roussot 1984.

131. Delporte 1973.

132. Roussot 1984.

133. Delluc & Delluc 1978, p. 389; a picture of the Cap Blanc tools can be seen, though with no indication of size, in White 1986, p. 135.

134. Roussot 1981; Laming-Emperaire 1962, p. 186. See also Simek 1986, p. 406.

135. Regnault 1906, p. 332.

136. Bader 1965, p. 28; Cârciumaru & Bitiri 1983. For a list of pigment analyses since 1950 (mostly at Lascaux & Altamira) see Couraud 1983, p. 108.

137. For Niaux, see Brunet 1981, 1982; for Las Monedas, see Ripoll 1972, p. 53.

138. Judson 1959.

139. Couraud 1982, p. 4.

140. Couraud 1983, p. 107; 1983a, p. 6.

141. Couraud & Laming-Emperaire, in Leroi-Gourhan & Allain 1979, pp. 153-70.

142. Couraud 1982.

143. Ballet *et al* in Leroi-Gourhan & Allain 1979, pp. 171-4.

144. Vandiver (in press).

145. Couraud 1982, p. 4; and in Leroi-Gourhan & Allain 1979, pp. 162-4. Couraud's experiments thus confirmed the opinions of Rottländer (1965).

146. For Baume-Latrone see *L'Art des Cavernes* 1984, pp. 333-9, and Bednarik 1986, p. 36; for Pileta, see Dams 1978, pp. 86/7.

147. Marshack 1985, pp. 78, 81.

148. Perpère 1984, p. 43.

149. Couraud 1982, p. 4; and in Leroi-Gourhan & Allain 1979, p. 165.

150. Leroi-Gourhan 1981, p. 30; 1982, p. 13.

151. Plenier 1971, pp. 69/70.

152. Marshack 1985, pp. 77-9.

153. Bohigas *et al* 1986; Moure Romanillo *et al* 1984/5; Almagro 1969.

154. Couraud 1982, p. 4; Barrière 1976, p. 75; Lorblanchet 1980, p. 34.

155. Pfeiffer 1982, pp. 241/2; Wildgoose *et al* 1982.

156. Leroi-Gourhan 1967. See also Moore 1977; Walsh 1979; Wright 1985.

157. Lorblanchet 1980, pp. 35, 37; Pfeiffer 1982, pp. 241/3.

158. Couraud 1982, p. 4. The experiments reported by Wildgoose *et al* produced somewhat equivocal results on this point and others.

159. Sahly 1963; Barrière & Sahly 1964. The medical view of the Gargas hands was presented earlier by Janssens 1957. For a review of incomplete hand stencils, see Pradel 1975.

160. Lorblanchet 1980, pp. 37/8.

161. Couraud 1982, p. 5.

162. Glory 1964; Couraud 1982, p. 5; and in Leroi-Gourhan & Allain 1979, p. 166.

163. Marshack 1985, p. 68.

164. J. Bégouën 1939, p. 284.

165. Altuna & Apellániz 1976.

166. Plenier 1971, pp. 69/70; de Balbín Behrmann & Moure Romanillo 1982, pp. 62/3.

167. Aujoulat 1985; Pfeiffer 1982, p. 142.

168. Rouzaud 1978, pp. 126/7.

169. Breuil & Lantier 1959, p. 223.

170. Sauvet 1983, p. 52.

171. For Roucadour, see *L'Art des Cavernes* 1984, pp. 511-13; for Baume-Latrone, see Bednarik 1986, p. 34.

172. Delluc & Delluc, in Leroi-Gourhan & Allain 1979, pp. 175-84; Barrière & Sahly 1964, p. 179.

173. *L'Art des Cavernes* 1984, pp. 596-9.

174. de Beaune 1987, 1987a.

175. de Beaune 1987, 1987a, 1984; Delluc & Delluc, in Leroi-Gourhan & Allain 1979, pp. 121-42; Ruspoli 1987, pp. 28-30.

176. Roussot 1984, p. 76.

177. de Beaune 1987, 1987a, 1984.

178. Fortea 1981.

179. Martín Santamaría & Moure Romanillo 1981. For La Griega, see Sauvet 1983.

180. Jorge *et al* 1981, 1982.
181. Sacchi, Abelanet & Brulé 1987; Bahn 1985.
182. Omnès 1982, pp. 105-7.

6. *What was Depicted?*

1. Examples cited in Bahn 1985a, from pp. 166, 283/284 and 328 of Bandi *et al* 1984.
2. See Rosenfeld 1984, and Macintosh 1977; Macintosh compared his identifications with the answers later given by an Aboriginal informant, and found that out of 22 items he had been wrong about 15, and only superficially right about the other 7!
3. Cited by Clottes 1986, p. 24.
4. Examples given by Clottes 1986, p. 10, and by Pales 1969, p. 52.
5. Rousseau adopted 'valable' and 'probable' categories in his 1967 study of felines, but Pales (1969, p. 52) disagreed with some of his assessments, finding him too ready to accept examples into both groups.
6. Pales 1969; Pales & de St Péreuse 1976, 1981, & forthcoming.
7. Lorblanchet 1977.
8. Leroi-Gourhan 1983, p. 258.
9. Leroi-Gourhan 1982, p. 49; 1983, p. 258.
10. Pales & de St Péreuse 1981, Nos 66/7, pls 84-7; 1976, pl. 52.
11. Rousseau 1967, p. 76; 1974; Pales 1969, p. 64.
12. Lhote 1968; Leroi-Gourhan 1958b, pp. 520/1; for examples of headless animals, see Leroi-Gourhan 1971, p. 471; and Ripoll 1972, pp. 56-60.
13. G., L. & R. Simonnet 1984, p. 29; also illustrated in Clottes 1986.
14. Bahn & Cole 1986, p. 144 & Fig. 28.
15. Lorblanchet 1974, p. 70; 1977, p. 46.
16. Bahn 1986, p. 117, & p. 107 (comment by Clottes); Davis 1986, p. 516.
17. Ucko & Layton 1984.
18. Rousseau 1984a; Barandiarán 1972, pp. 351-8.
19. Guthrie 1984, p. 38; Mithen (1988) prefers to see these features as teaching aids for hunters.
20. For Morin, see de Sonneville-Bordes 1986, p. 632; for the fawns, see Lorblanchet in Marshack 1985, pp. 62/3; for the Mas d'Azil bone disc, see Chollot-Varagnac 1980, p. 408 (many scholars, including Graziosi 1960, Barandiarán 1968, Leroi-Gourhan 1971 and Apellániz 1982, 1986, have wrongly attributed this piece to Laugerie-Basse).
21. See examples in Lhote 1968.
22. Baffier 1984, pp. 143, 150.
23. Baffier 1984.
24. Pales & de St Péreuse 1981, p. 87.
25. Leason 1939, 1956; Souriau 1971 likewise believes that the deer on the Lortet baton are stretched dead on the ground, one with head bent back, whereas most scholars have seen them as galloping and bellowing.
26. Riddell 1940.
27. Pales & de St Péreuse 1965, p. 229; 1981, p. 96; Pales 1969, p. 115.
28. Guthrie 1984, p. 46; Riddell 1940, p. 160.
29. Vojkffy, quoted in Bandi 1968, p. 16; Bandi *et al* 1984, p. 29.
30. Vojkffy, quoted in Giedion 1965, p. 193.
31. Bandi 1968, p. 17; Breuil & Obermaier 1935, p. 31 and pl. 27; Leroi-Gourhan 1984, p. 77; see also Jordá 1980, p. 282.
32. Baffier 1984, p. 153.
33. Leroi-Gourhan 1984, p. 76.
34. Graziosi 1960, pp. 91/2 and pl. 90; Barandiarán 1971a; 1973, pp. 187/8 and pl. 45, No. 1; Breuil 1936/7, p. 6; Bahn 1982, pp. 257/8. Three of the 'wolf/cervid' pieces were also studied in Sieveking 1978.
35. Delporte 1984, pp. 112-14; see also Caralp *et al* 1973.
36. For a general survey of the fauna and its depictions, see Powers & Stringer 1975.
37. Leroi-Gourhan 1971, p. 463.
38. For example, Piette 1887.
39. Capitan, Breuil & Peyrony 1924; Bourdelle 1956; Blanchard 1964; for a summary of the problems, see Barandiarán 1972, p. 346; Pales & de St Péreuse 1981, p. 58; Prat 1986; and de Sonneville-Bordes & Laurent 1986/7, p. 72.
40. Lión 1971; Altuna & Apellániz 1978, pp. 107-9, 113-19; Bosinski & Fischer 1980; Pales & de St Péreuse 1981.
41. Pales & de St Péreuse 1981, pp. 26/7.
42. Pales & de St Péreuse 1981, p. 38.
43. Leroi-Gourhan 1984, pp. 76-8.
44. Schmid 1984, p. 157.
45. Altuna 1975, pp. 112/3; Altuna & Apellániz 1978; see also Barandiarán 1972, pp. 365-8; and Bahn 1982a, p. 23.
46. Rousseau 1973; also 1984a, pp. 187-9; Altuna & Apellániz 1978; and Barandiarán 1972, p. 373.
47. Bahn 1982a, p. 21.
48. Bahn 1983a.
49. Arambourou 1978, p. 49; Bahn 1982a, pp. 22/3.
50. Piette 1906; Pales & de St Péreuse 1966, & 1981, pl. 1; Bahn 1978a; Jordá 1987.
51. Züchner 1975, p. 23; for an earlier study of bison figures, see Capitan, Breuil & Peyrony 1910.
52. Pales & de St Péreuse 1981, pp. 63-106; 1965.
53. de Lumley 1968a; for objections, see Pales & de St Péreuse 1981, pp. 75-8.

54. Pales & de St Péreuse 1981, p. 72.
55. Leroi-Gourhan 1984, pp. 77-80.
56. It was primarily Breuil & Laming-Emperaire who said the small bovids were *Bos longifrons;* the sexual dimorphism solution was put forward by Zeuner (1953) and Koby (1954).
57. Koby 1968.
58. Barandiarán 1969; Ripoll 1972, & 1984, p. 272; see also Olivié 1984/5 for cervids in Cantabrian art.
59. Faure 1978, pp. 47, 50, 76.
60. Vialou 1984, and in Leroi-Gourhan & Allain 1979.
61. Kehoe 1987 & in press; and letter to T.F. Kehoe from Danish zoologist M. Meldgaard, 30/3/87. Other possible examples of 'swimming' animals may include the deer on the Lortet baton, with fish between their legs and 'above' their backs; and some horses at Lascaux are said to hold their heads just as if they were crossing a river (according to Gunn, in Marshack 1979, p. 299).
62. Blanchard 1964a.
63. Vialou 1984, p. 214.
64. Bouchud 1966.
65. Pales & de St Péreuse, forthcoming.
66. Schmid, in Bandi *et al* 1984, p. 90.
67. Leroi-Gourhan 1984, pp. 81, 84; Guthrie 1984, p. 45.
68. de Sonneville-Bordes 1986, p. 633.
69. Pales 1976/7; Pales & de St Péreuse 1981, pp. 107-40.
70. Pales 1976/7, pp. 90/91; Pales & de St Péreuse 1981, pp. 128-31.
71. Pales 1976/7, pp. 91, 101/2; Pales & de St Péreuse 1981, pp. 132/3.
72. Pales & de St Péreuse 1981, pp. 137-9. For an uncritical review of ibex figures in Pyrenean art, see Brielle 1968.
73. Welté 1975/6.
74. Pales & de St Péreuse, forthcoming.
75. Bosinski 1984; Bosinski & Fischer 1980. For an earlier, less critical review of mammoths in art, see Berdin 1970.
76. Bosinski 1984, p. 301; Bosinski & Fischer 1980, pp. 38-40; Hahn 1984, p. 288; Praslov 1985, p. 186.
77. Ripoll 1984, p. 278; Bandi 1968, p. 15.
78. Welté 1975/6; Leroi-Gourhan 1984, p. 83; for Pech Merle, see Lorblanchet 1977, pp. 54/5.
79. Jordá 1983.
80. Capitan *et al* 1924; Breuil, Nougier & Robert 1956; Nougier & Robert 1965, 1966.
81. Rousseau 1967, 1974; Pales 1969; Altuna 1972, pp. 308-15.
82. Pales 1969, p. 102.
83. Leroi-Gourhan 1984, p. 84; Guthrie 1984, p. 44.
84. Klíma 1984, p. 326.
85. Pales 1969, pp. 53, 84, 102.
86. Mazak 1970; Rousseau 1974.
87. Pales 1969, pp. 58-60; Rousseau 1984a, p. 170.
88. For general studies of rare species, see Capitan *et al* 1910, 1924; Novel 1986; for examples of studies (both critical and uncritical) of individual species, see Barandiarán 1974 (glutton); Nougier & Robert 1957a (rhino), 1958 (saïga), and 1960 (canids).
89. Capitan *et al* 1924; de Sonneville-Bordes & Laurent 1983; de Sonneville-Bordes 1986, pp. 639-41; Bahn 1977, pp. 253/4; 1982, p. 255; 1984.
90. Marshack 1970, 1972, 1975. For the whale theory, see Robineau 1984.
91. Dams 1987, p. 17; 1987a, pp. 219-23.
92. Breuil & de St Périer 1927.
93. Bahn 1984, pp. 153, 157, 243-5, 276-8; Marshack 1970, 1972. For other features of fish depictions, see Barandiarán 1972, pp. 348-51.
94. Dams 1987; 1978, p. 85.
95. Alcalde, Breuil & Sierra 1912, ch. XVI; Vayson de Pradenne 1934; Breuil & Bégouën 1937; Lorblanchet 1974, pp. 85-96.
96. Buisson & Pinçon 1986/7, p. 79.
97. Eastham 1986.
98. Bahn & Butlin 1987.
99. Tyldesley & Bahn 1983; Delcourt-Vlaeminck 1975; Marshack 1972, 1975.
100. Leroi-Gourhan 1971.
101. Conkey 1981.
102. Sauvet 1979, pp. 342/3; Leroi-Gourhan 1971, Fig. 764; Vialou 1986.
103. Nougier 1972, p. 265.
104. Roussot 1984b. For the statistical problems involved in making cumulative percentage frequency graphs to compare assemblages ranging from hundreds of figures to fewer than ten, see J.E. Kerrich & D.L. Clarke, 'Notes on the possible misuse and errors of cumulative percentage frequency graphs for the comparison of prehistoric artefact assemblages', in *Proc. Prehist. Soc.* 33, 1967, pp. 91-116.
105. Delporte 1984, p. 132; Capitan & Peyrony 1928, pp. 107-15.
106. Bosinski 1984, p. 320; Bosinski & Fischer 1980, pp. 56-60, 124-6.
107. Ucko & Rosenfeld 1972, p. 156. For an early study of humans in the art, see Capitan *et al* 1924, pp. 91-116.
108. Macintosh 1977, p. 191. See also Moore 1977; Walsh 1979; Wright 1985.

109. Pales & de St Péreuse 1976; 1981, p. 28; Airvaux & Pradel 1984.
110. Brodrick 1963, pp. 126/7.
111. Bahn 1986.
112. Pales & de St Péreuse 1976, p. 30; Airvaux & Pradel 1984.
113. Pales & de St Péreuse 1976, pp. 31-4.
114. Delporte 1979, p. 27.
115. Leroi-Gourhan 1971, Fig. 777; see Pales & de St Péreuse 1976, p. 49; Ucko & Rosenfeld 1972, pp. 194.
116. de Saint Mathurin 1973; 1975, pp. 25/6.
117. Pales & de St Péreuse 1976, p. 156.
118. Pales & de St Péreuse 1976, pp. 83-5; Ucko & Rosenfeld 1972, pp. 181/2. The abundance of humans with arms raised was noticed by Cartailhac & Breuil 1906, p. 52.
119. Pales & de St Péreuse 1976, p. 52.
120. Pales & de St Péreuse 1976, pp. 28/9.
121. Out of a vast literature, see Delporte 1979; Passemard 1938; Narr 1960; Abramova 1967.
122. Pales 1972, p. 218.
123. Piette 1895, 1902; Ucko & Rosenfeld 1972, p. 161.
124. Pales 1972, p. 220; Pales & de St Péreuse 1976, pp. 91/2, 32.
125. Rice 1981; see also Guthrie 1984, p. 62; the Rice paper is less useful when it compares its results to the supposed age structure of Upper Palaeolithic adult females, since it has lumped together figurines spanning the entire period.
126. Leroi-Gourhan 1971, Fig. 52 bis; Pales 1972; Pales & de St Péreuse 1976, pp. 70-74. See also Delporte 1971; 1979, pp. 264-8.
127. Delporte 1962, p. 58.
128. de Saint Mathurin 1978, 1973.
129. Abramova 1967; 1984, p. 333; Shimkin 1978, p. 278; Berenguer 1986, p. 668.
130. Praslov 1985, pp. 182/3.
131. Delporte 1979, p. 76; Gamble 1982, p. 98.
132. Delluc 1981a, p. 72.
133. Breuil & Lantier 1959, p. 237; Pales & de St Péreuse 1976, p. 101.
134. Carcauzon 1984, 1988.
135. Bosinski & Fischer 1974; Bosinski 1973.
136. Rosenfeld 1977; Lorblanchet & Welté 1987.
137. Bosinski & Fischer 1974, p. 45 & pl. 59.
138. Bosinski & Fischer 1974, p. 94.
139. Breuil 1905; Breuil & de St Périer 1927; Graziosi 1960, p. 196 and pl. 291; Leroi-Gourhan 1958a, p. 388; 1971, p. 480, Fig. 792.
140. Rosenfeld 1977, p. 101.
141. Ucko & Rosenfeld 1972, p. 168; Lorblanchet 1977. On claviforms, see also Stoliar 1977/8a, p. 64.
142. Abramova 1967; Praslov 1985, p. 190; Stoliar 1977/8, pp. 11/12; 1977/8a, pp. 51-9, 73; Shimkin 1978, p. 279. See also Rosenfeld 1977, p. 93.
143. Rousseau 1967, p. 158/9. On masks, see also chapter in Cartailhac & Breuil 1906.
144. Giedion 1965, p. 374.
145. Ucko & Rosenfeld 1967; 1972, p. 205.
146. Bégouën & Breuil 1958; Rousseau 1967, p. 102.
147. Leroi-Gourhan 1983.
148. Pales & de St Péreuse 1976, p. 144.
149. Bosinski 1984, p. 318; Bosinski & Fischer 1974, 1980.
150. Conkey 1981, p. 24.
151. For examples, see Jordá 1979, 1985.
152. Lorblanchet 1977; Bahn 1986, p. 104.
153. See quotations from Forge, Munn and Warner, in Bahn 1986, pp. 104, 109. See also Speck & Schaeffer 1950.
154. And. Leroi-Gourhan 1980; Sauvet & Wlodarczyk 1977, pp. 548-50; Casado 1977.
155. Capdeville 1986; Vialou 1987; there is also a possible tectiform engraved on a bone tool from Altamira.
156. Leroi-Gourhan 1958, 1958a, 1968.
157. Casado 1977; Sauvet & Wlodarczyk 1977.
158. Mons 1980/1; Chollot-Varagnac (1980) established 25 basic motifs.
159. Conkey 1980, 1981.
160. Sauvet & Wlodarczyk 1977, p. 553.
161. eg see Piette 1905.
162. Forbes & Crowder 1979.
163. Sauvet & Wlodarczyk 1977, p. 552.
164. See, for example, Clegg 1985, 1986.
165. Casado 1977, p. 251.
166. Leroi-Gourhan 1958, p. 318; 1971; Sauvet & Wlodarczyk 1977, p. 557.
167. Geoffroy 1974, p. 47.
168. Geoffroy 1974, p. 57.
169. González García 1985, 1987; Geoffroy 1974, p. 55.

7. *Reading the Messages*

1. For an account of Piette's view, see Pales 1969, p. 111. On de Mortillet, see Reinach 1899; Bahn 1988; Laming-Emperaire 1962, p. 70, and Ucko & Rosenfeld 1967, p. 118.
2. Quoted in Laming-Emperaire 1962, p. 66; and Pales 1969, p. 112.
3. Halverson 1987; this paper also provides part of the history of the theory. See also Souriau 1971.
4. Chollot-Varagnac 1980, p. 448.
5. Lewis-Williams 1982, p. 429; Bahn 1987a.
6. Reinach 1903; see also Laming-Emperaire 1962, pp. 72-5.

7. Cartailhac & Breuil 1906 (see especially pp. 143 and 225).
8. Breuil 1952; Bégouën 1929.
9. Giedion 1965, p. 57.
10. Lips 1949, pp. 84/5; Lindner 1950, pp. 53-67.
11. Obermaier 1918.
12. Graziosi 1960, p. 152. On the Montespan case, and other fallacious evidence for hunting and magic, see Bahn 1988a.
13. Pales 1969, p. 47.
14. Leroi-Gourhan 1958a, p. 390; 1971, p. 30.
15. Faure 1978, p. 61.
16. Faure 1978.
17. Barandiarán 1984.
18. Guthrie 1984, p. 51, states wrongly that 'portrayals of hunting scenes constitute the bulk of Palaeolithic large animal art'. Similarly, Mithen (1988) has interpreted all manner of figures and shapes as examples of tracks, droppings, and other visual 'cues' in an attempt to see the art as a means of communicating information about hunting.
19. Guthrie 1984, p. 50.
20. Eaton 1978.
21. Parkington 1969, p. 12. See Rice & Paterson 1986, p. 665, for a similar suggestion.
22. eg Breuil & Lantier 1959, pp. 237/8; Lindner 1950, p. 69; Marshack 1976b, pp. 71/2.
23. Lindner 1950, pp. 56/7.
24. Kehoe 1987, in press.
25. Weissen-Szumlanska 1951, pp. 457/8.
26. Kehoe, in press.
27. Pales & de St Péreuse 1976, p. 78.
28. eg Chollot-Legoux 1961.
29. Delluc 1985, 1981b; Breuil 1925.
30. Bahn 1986; Delluc 1985,
31. Bosinski 1973, p. 45; 1984, p. 315.
32. Klíma 1984, pp. 324-6.
33. Delporte 1984, p. 132.
34. de Sonneville-Bordes & Laurent 1986/7, pp. 69/70.
35. Delluc 1984c; 1981a.
36. Altuna 1983; Altuna & Apellániz 1978, pp. 106-7.
37. Delporte 1984, p. 125.
38. Roussot 1984b, pp. 495/6.
39. Rice & Paterson 1985, 1986.
40. See for example Lommel 1967; Davenport & Jochim 1988.
41. Glory 1968, pp. 37/8, 57.
42. Smith, in press.
43. Clark 1966, pp. 12/13.
44. Lewis-Williams 1982; Lewis-Williams & Dowson 1988.
45. Vialou 1986, p. 139.
46. Laming-Emperaire 1962, p. 219/20 & pl. 13; Leroi-Gourhan 1958a, p. 396; 1966, p. 41; 1971, p. 387, Figs. 491-3; Bouvier 1976.
47. Freeman 1978; 1984, p. 214; but see Breuil & Obermaier 1935, pp. 85/6, who wisely leave the second animal undetermined.
48. Freeman 1978; 1984, p. 222,
49. Nougier & Robert 1974.
50. J.R.B. Speed, Vet.M.B., M.R.C.V.S., pers. comm.
51. Leroi-Gourhan 1966, p. 39; 1971, pp. 89, 98; Bandi 1968, p. 16; Baffier 1984, pp. 148/9.
52. Glory 1968, p. 57; Rousseau 1984a, p. 195.
53. Pales & de St Péreuse 1976, pp. 114-23.
54. Bégouën et al 1982, 1984, 1984/5 pp. 66-70.
55. Cabré 1934.
56. L. Bégouën 1939, pp. 293/4; Bégouën & Clottes 1984; Bégouën et al 1984/5, pp. 29-33. A further example of such wishful thinking is provided by a small limestone sculpture from Laussel which seems to be the gland of a phallus; Lalanne claimed in 1946 that it came from a carved erection, but in fact this cannot be deduced from a gland alone (see Duhard & Roussot 1988, p. 43).
57. St Périer 1936, p. 115.
58. Rice 1981.
59. On Kostenki, Praslov 1985, p. 185; on Tursac, Delporte 1979, p. 76; on Grimaldi, Duhard 1987; on Monpazier, Duhard 1987a.
60. Leroi-Gourhan 1971, p. 480, Fig. 794; Bahn 1986, pp. 109, 119.
61. Delluc 1978, p. 239; Stoliar 1977/8a, p. 42; Bahn 1986, p. 99.
62. Bahn 1986.
63. Collins & Onians 1978; Delluc 1978, p. 353; and see Bahn 1986, p. 102.
64. Stoliar 1977/8a, p. 39.
65. For the phallus theory, see Bégouën & Breuil 1958, p. 98, and Vialou 1986, p. 198; for the horn theory, Ucko & Rosenfeld 1967, pp. 57 & 178; for the sculptor's view, Beasley 1986.
66. de Saint Mathurin 1978, p. 17.
67. Bahn 1986, pp. 107, 117; Delluc 1985, pp. 56/7, 61.
68. For illustrations, see Jelínek 1975, pp. 406-10.
69. Guthrie 1984, pp. 62-6.
70. Collins & Onians 1978.
71. Guthrie 1984, pp. 70/1.
72. Guthrie 1984 made the comparison; it was adopted by Kurtén 1986, from whom the quotation is taken.
73. Bahn 1986, 1985a; Gimbutas 1981.
74. Cartailhac 1902, p. 349.

75. Raphael 1986; Laming-Emperaire 1962; for Leroi-Gourhan, see especially 1971, 1982. Two collections of his papers have been published in Spanish: 1984a has his major articles and bibliography from 1935 to 1983; 1984b has his lectures at the Collège de France from 1969 to 1983. A critical review of the development of his views on Palaeolithic art can be found in Meylan 1986, and a critique of his early work in Ucko & Rosenfeld 1967, pp. 195-221.
76. Leroi-Gourhan 1982, pp. 45-50.
77. Leroi-Gourhan 1958a, p. 395; 1971.
78. Leroi-Gourhan 1958, p. 321.
79. Leroi-Gourhan 1982; 1972, p. 285.
80. Leroi-Gourhan 1972, p. 307.
81. Jordá 1979, 1985.
82. Leroi-Gourhan 1971, pp. 86, 322; 1972, p. 308.
83. Parkington 1969, pp. 5/6; Ucko & Rosenfeld 1967; Bandi 1972, p. 315; Lhote 1972, p. 322.
84. Meylan 1986, pp. 45/6.
85. Leroi-Gourhan 1972, p. 291.
86. Leroi-Gourhan 1982, p. 50.
87. eg Altuna & Apellániz 1976, p. 163; 1978; Jordá 1980; Vialou 1986.
88. Leroi-Gourhan 1972, p. 306; 1968.
89. Leroi-Gourhan 1972, p. 289.
90. Leroi-Gourhan 1971, p. 481.
91. Lhote 1968a; 1972, pp. 324-8.
92. Leroi-Gourhan 1966, p. 44.
93. Leroi-Gourhan 1958a; 1968. On the La Madeleine 'phallus', see Bahn 1986 pp. 103, 114, 118.
94. Breuil & Lantier 1959, p. 237.
95. Leroi-Gourhan 1982, p. 58; see also Bahn 1978b, 1980.
96. Leroi-Gourhan 1971, p. 259; but see Ucko & Rosenfeld 1967, pp. 195-221; 1972, pp. 177, 193-7.
97. Leroi-Gourhan 1971, pp. 95/6; Pales & de St Péreuse 1976, pp. 49, 154.
98. Pales & de St Péreuse 1976, pp. 153-5.
99. Pales & de St Péreuse 1976, p. 27; 1981, p. 28; for other examples of associations in Palaeolithic art, see Nougier & Robert 1968, Caralp et al 1973.
100. Sauvet 1979, p. 347.
101. Sauvet 1979, p. 348.
102. Parkington 1969. For a different statistical test, see Ucko & Rosenfeld 1967, p. 208.
103. Stevens 1975, 1975a.
104. Laming-Emperaire 1962, pp. 115-23. On Raphael, see pp. 118/9; see also Raphael 1986.
105. Laming-Emperaire 1972, pp. 66/7.
106. Laming-Emperaire 1970; 1971; 1972, p. 70.
107. González García 1987; 1985, p. 490.
108. González García 1985, pp. 517-19.
109. Rivenq, pers. comm.; Bégouën 1984, p. 77.
110. González García 1985, pp. 457-63; Eastham 1979, pp. 370/71.
111. Nougier 1975; 1975a, p. 115.
112. L'Art des Cavernes 1984, pp. 544-8.
113. Bahn 1978b, 1980.
114. Jochim 1983.
115. Bahn 1984; Mellars 1985, p. 283.
116. Sieveking 1979a, p. 106, supports the former view; against it, see Bahn 1977, p. 251; 1978b, p. 125; and Conkey 1983, pp. 206, 221.
117. Lorblanchet 1980; and comment in Marshack 1985, p. 64.
118. Nougier, eg 1972.
119. Pales & de St Péreuse 1965, p. 230; Pales 1969, pp. 110/11.
120. Apellániz 1982, pp. 38/9; Mons 1986/7, p. 91.
121. Almagro 1976, p. 71; Bosinski 1973, p. 43; 1984, p. 274.
122. Breuil 1962, p. 356; for early views on Altamira, see Apellániz 1982, p. 48; 1983, p. 274; Cartailhac & Breuil 1906.
123. Leroi-Gourhan 1971, p. 29.
124. Pales & de St Péreuse 1981, p. 139; see also Baffier 1984, p. 153.
125. Apellániz 1980.
126. Apellániz 1982, pp. 45-63; 1983.
127. Apellániz 1982, pp. 34-8; 1986, pp. 50-52. It will be recalled that, like other authors before him, he attributes the Mas d'Azil disc wrongly to Laugerie-Basse, and that of Laugerie-Basse to abri du Souci.
128. Altuna & Apellániz 1976, pp. 148-53; 1978, p. 141; Apellániz 1982.
129. For Chimeneas, see Apellániz 1982, pp. 63-8; 1984. For Lascaux, see 1984a.
130. Altuna & Apellániz 1978, pp. 125-33; Apellániz 1982, pp. 71-92; Arte Rupestre 1987, p. 40.
131. Sauvet 1983, pp. 56/7; but see reply by Apellániz in Ars Praehistorica 3/4, 1984/5, pp. 259-60.
132. Apellániz 1984, pp. 534-7; 1987.
133. Bahn 1982, 1984; Conkey 1980; Soffer 1985.
134. Breuil 1952, pp. 194/5; Leroi-Gourhan 1968, p. 69; Casado 1977, p. 16.
135. Leroi-Gourhan 1966, p. 47.
136. Eastham 1979, pp. 378-84.
137. For Limeuil see Capitan & Bouyssonie 1924; for Mezhirich, see Gladkih et al 1984, p. 141; Shimkin 1978, pp. 274, 283; Soffer 1985, p. 79.
138. Marshack 1979, pp. 287-92.
139. Dewez 1974.
140. Frolov 1977/9; 1981; and in Marshack 1979, pp. 605-7.

141. Marshack 1979, pp. 271, 309, 607; 1972a, p. 329.
142. Marshack 1970a, 1972, 1972a, 1975. See also review by Rosenfeld in *Antiquity* 45, 1971, pp. 317-19, and reply in 46, 1972, pp. 63-5.
143. Couraud 1985; Bahn & Couraud 1984.
144. Frolov 1977/9; 1979; 1981.
145. Guthrie 1984, p. 47; Marshack 1970, 1972, 1975; but see Faure 1978, Bahn 1986.
146. Marshack 1977; 1979, p. 305.
147. Marshack 1985, pp. 70, 78-81.
148. Marshack 1969, 1972, 1985.
149. Lorblanchet 1980a, p. 476; and in Marshack 1985, pp. 62/3.
150. Leroi-Gourhan 1981, pp. 25, 32; Lorblanchet 1980a, pp. 474, 476.
151. Marshack 1984, 1985.
152. Collot *et al* 1982.
153. Pfeiffer 1982.
154. Pfeiffer 1982.
155. Yates 1966; Pfeiffer 1982, pp. 210-25.
156. Sauvet 1979.
157. Conkey 1978, 1980, 1985.
158. Conkey 1980.
159. Gamble 1982, p. 104.
160. Conkey 1983, pp. 218-19; 1984, p. 264.
161. Hammond 1974.
162. Bednarik 1986; Davis 1986.
163. Lewis-Williams & Dowson 1988.
164. Comment by Bahn, in Lewis-Williams & Dowson 1988, pp. 217/8.
165. Lewis-Williams 1982; Lewis-Williams & Dowson 1988.
166. Lewis-Williams 1982, p. 434.
167. Pericot 1962; Bahn 1986, pp. 110, 119.
168. Forbes & Crowder 1979, p. 362 & plate.

Bibliography

This is by no means intended to be an exhaustive list of references on the subject of Palaeolithic art: instead it concentrates on recent studies. Extensive bibliographies covering the older literature, and also concerning individual sites, can be found in: Breuil 1952; Zervos 1959; Graziosi 1960; Laming-Emperaire 1962; Leroi-Gourhan 1971; Naber *et al* 1976; and *L'Art des Cavernes* 1984.

The most up-to-date guide to the few French decorated caves open to the public is Vialou 1976 (though it predates Lascaux II); *L'Art des Cavernes* also has some information on opening hours; the only equivalent for northern Spain, though now very outdated, is Sieveking & Sieveking 1962.

ABRAMOVA, Z.A. 1967. Palaeolithic art in the USSR. *Arctic Anthropology* 4 (2), 1-179. Also 'L'art mobilier paléolithique en URSS', in *Quartär* 18, 1967, 99-125, 9 pl.

ABRAMOVA, Z.A. 1984. Les corrélations entre l'art et la faune dans le Paléolithique de la Plaine russe (La femme et le mammouth), in Bandi *et al* 1984, 333-42.

ABSOLON, C. 1937. Les flûtes paléolithiques de l'Aurignacien et du Magdalénien de Moravie (analyse musicale et ethnologique comparative, avec démonstrations). *12e Congrès Préhist. de France*, Toulouse/Foix 1936, 770-84.

ADAM, K.D. & KURZ, R. 1980. *Eiszeitkunst im süddeutschen Raum*. Konrad Theiss Verlag: Stuttgart.

AIKENS, C.M. & HIGUCHI, T. 1982. *Prehistory of Japan*. Academic Press: London/New York.

AIRVAUX, J. & PRADEL, L. 1984. Gravure d'une tête humaine de face dans le Magdalénien III de La Marche, commune de Lussac-les-Châteaux (Vienne). *Bull. Soc. Préhist. française* 81, 212-15.

AIRVAUX, J. *et al* 1983. La plaquette gravée du Périgordien supérieur de l'abri Laraux, commune de Lussac-les-Châteaux (Vienne). Nouvelle lecture et comparaisons. *Bull. Soc. Préhist. française* 80, 235-46.

ALCALDE DEL RIO, H., BREUIL, H. & SIERRA, L. 1912. *Les Cavernes de la Région Cantabrique*. Monaco.

ALLAIN, J. 1950. Un appeau magdalénien. *Bull. Soc. Préhist. française* 47, 181-92.

ALLAIN, J. & RIGAUD, A. 1986. Décor et fonction. Quelques exemples tirés du Magdalénien. *L'Anthropologie* 90, 713-38.

ALMAGRO BASCH, M. 1969. *Las Pinturas Rupestres de la cueva de Maltravieso en Cáceres*. Min. de Educación y Ciencia: Madrid.

ALMAGRO BASCH, M. 1976. Los omoplatos decorados de la cueva de 'El Castillo', Puente Viesgo (Santander). *Trabajos de Prehistoria* 33, 9-99, 12 pl.

ALMAGRO BASCH, M. 1980. Los grabados de trazo múltiple en el arte cuaternario español, in *Altamira Symposium*, 27-71. Min. de Cultura: Madrid.

ALMAGRO BASCH, M. 1981. La tecnica del grabado de trazos múltiples en el arte cuaternario español, in *Arte Paleolítico*, Comisión XI, Xth Congress UISPP, Mexico City, 1-52.

ALMAGRO BASCH, M., GARCIA GUINEA, M.A. & BERENGUER, M. 1972. La época de las pinturas y esculturas polícromas cuaternarias en relación con los yacimientos de las cuevas: revalorización del Magdaleniense III. in *Santander Symposium* 467-73.

ALTUNA, J. 1972. Fauna de mamíferos de los yacimientos prehistóricos de Guipúzcoa. *Munibe* 24, 464pp.

ALTUNA, J. 1975. *Lehen euskal herria – Guide illustré de préhistoire basque.* Mensajero: Bilbao.

ALTUNA, J. 1983. On the relationship between archaeofaunas and parietal art in the caves of the Cantabrian region, in *Animals and Archaeology, I: Hunters and their Prey* (J. Clutton-Brock & C. Grigson, eds), 227-38. British Arch. Reports, Int. series 163, Oxford.

ALTUNA, J. & APELLANIZ, J.M. 1976. Las figuras paleolíticas de la cueva de Altxerri (Guipúzcoa). *Munibe* 28, 3-242.

ALTUNA, J. & APELLANIZ, J.M. 1978. Las figuras rupestres de la cueva de Ekain (Deva). *Munibe* 30, 1-151.

ANATI, E. 1985. The rock art of Tanzania and the East African sequence. *Bollettino del Centro Camuno di Studi Preistorici* 23, 15-68.

ANATI, E. 1986. Etat de la recherche sur l'art rupestre: rapport mondial. *L'Anthropologie* 90, 783-800.

ANDRIEUX, C. 1974. Premiers résultats sur l'étude du climat de la salle des peintures de la Galerie Clastres (Niaux, Ariège). *Annales de Spéléologie* 29, 3-25.

APELLANIZ, J.M. 1980. El método de determinación de autor en el Cantábrico. Los grabadores de Llonín, in *Altamira Symposium*, 73-84. Min. de Cultura: Madrid.

APELLANIZ, J.M. 1982. *El Arte Prehistórico del Pais Vasco y sus Vecinos.* Desclée de Brouwer: Bilbao.

APELLANIZ, J.M. 1983. El autor de los bisontes tumbados del techo de los polícromos de Altamira, in *Homenaje al Prof. M. Almagro Basch*, vol. 1, 273-80. Min. de Cultura: Madrid.

APELLANIZ, J.M. 1984. La méthode de détermination d'auteur appliquée à l'art pariétal paléolithique. L'auteur des cervidés à silhouette noire de Las Chimeneas (Santander, Espagne). *L'Anthropologie* 88, 531-37.

APELLANIZ, J.M. 1984a. L'auteur des grands taureaux de Lascaux et ses successeurs. *Ibid.* 539-61.

APELLANIZ, J.M. 1986. Análisis de la variación formal y la autoría en la iconografía mueble del Magdaleniense Antiguo de Bolinkoba (Vizcaya). *Munibe* 38, 39-59.

APELLANIZ, J.M. 1987. Aplicación de técnicas estadísticas al análisis iconográfico y al método de determinación de autor. *Munibe* 39, 39-60.

ARAMBOUROU, R. 1978. *Le Gisement Préhistorique de Duruthy à Sorde-l'Abbaye (Landes). Bilan des Recherches de 1958 à 1975.* Mémoire 13 de la Soc. Préhist. fr., 158pp.

L'ART DES CAVERNES 1984. *Atlas des Grottes Ornées Paléolithiques françaises.* Min. de la Culture: Paris.

ARTE RUPESTRE EN ESPANA 1987. *Revista de Arqueología*, numero extra.

ASLIN, G.D. & BEDNARIK, R.G. 1984a. Karliengoinpool Cave: A preliminary report. *Rock Art Research* 1 (1), 36-45.

ASLIN, G.D. & BEDNARIK, R.G. 1984b. Koorine Cave, South Australia. *Rock Art Research* 1 (2), 142-44.

ASLIN, G.D., BEDNARIK, E.K. & BEDNARIK, R.G. 1985. The 'Parietal Markings Project' – a progress report. *Rock Art Research* 2 (1), 71-74.

ASLIN, G.D. & BEDNARIK, R.G. 1985. Mooraa Cave – a preliminary report. *Rock Art Research* 2 (2), 160-65.

ATLAS OF PRIMITIVE MAN IN CHINA 1980. Science Press: Beijing.

AUDOUIN, F. & PLISSON, H. 1982. Les ocres et leurs témoins au Paléolithique en France: enquête et expériences sur leur validité archéologique. *Cahiers du Centre de Recherches Préhistoriques* 8, 33-80.

AUJOULAT, N. 1985. Analyse d'une oeuvre pariétale anamorphosée. *Bull. Soc. Préhist. Ariège* 40, 185-93.

AUJOULAT, N. 1987. *Le Relevé des Oeuvres Pariétales Paléolithiques. Enregistrement et Traitement des Données.* Documents d'Archéologie française No. 9. Maison des Sciences de l'Homme: Paris.

AUJOULAT, N., ROUSSOT, A. & RIGAUD, J-P. 1984. Traces peu explicites et attributions douteuses ou erronées, in *L'Art des Cavernes*, 72-85. Min. de Culture: Paris.

AVELEYRA ARROYO DE ANDA, L. 1965. The Pleistocene carved bone from Tequixquiac, Mexico: A reappraisal. *American Antiquity* 30, 261-77.

BADER, O.N. 1965. *La Caverne Kapovaïa, Peinture Paléolithique.* Nauka: Moscow.

BADER, O.N. 1967. Die paläolithischen Höhlenmalereien in Osteuropa. *Quartär* 18, 127-38.

BADER, O.N. 1976. Upper Palaeolithic Burials and the Grave at the Sungir Site. *Sovetskaia Arkheologiia* 3, 1967, 142-59, translated for the U.S. Geological Survey.

BAFFIER, D. 1984. Les caractères sexuels secondaires des mammifères dans l'art pariétal paléolithique franco-cantabrique, in Bandi *et al* 1984, 143-54.

BAHN, P.G. 1977. Seasonal migration in S.W. France during the late glacial period. *Journal of Arch. Science* 4, 245-57.

BAHN, P.G. 1978. Palaeolithic pottery, the history of an anomaly. *Anthropos* (Athens), 5, 98-110.

BAHN, P.G. 1978a. The 'unacceptable face' of the West European Upper Palaeolithic. *Antiquity* 52, 183-92, 1 pl.

BAHN, P.G. 1978b. Water mythology and the distribution of Palaeolithic parietal art. *Proc. Prehist. Soc.* 44, 125-34.

BAHN, P.G. 1980. 'Histoire d'Eau': L'art pariétal préhistorique des Pyrénées. *Travaux de l'Inst. d'Art Préhist. Toulouse* 22, 129-35.

BAHN, P.G. 1982. Inter-site and inter-regional links during the Upper Palaeolithic: the Pyrenean evidence. *The Oxford Journal of Arch.* 1, 247-68.

BAHN, P.G. 1982a. Homme et cheval dans le Quaternaire des Pays de l'Adour, in *Les Pays de l'Adour, Royaume du Cheval*, Guide-Catalogue, Musée Pyrénéen, Lourdes, pp. 21-26.

BAHN, P.G. 1983. A Palaeolithic treasure-house in the Pyrenees. *Nature* 302, 571-72.

BAHN, P.G. 1983a. New finds at Pincevent. *Nature* 304, 682-83.

BAHN, P.G. 1984. *Pyrenean Prehistory.* Aris & Phillips: Warminster.

BAHN, P.G. 1984a. How to spot a fake azilian pebble. *Nature* 308, p. 229.

BAHN, P.G. 1985. Ice Age drawings on open rock faces in the Pyrenees. *Nature* 313, 530-31.

BAHN, P.G. 1985a. Review of Bandi *et al* 1984. *Antiquity* 59, 57-58.

BAHN, P.G. 1986. No sex, please, we're Aurignacians. *Rock Art Research* 3, 99-120.

BAHN, P.G. 1987. A la recherche de l'iconographie paléolithique hors de l'Europe. *Travaux de l'Institut d'Art Préhist. de Toulouse* 29, 7-18.

BAHN, P.G. 1987a. Comment on article by J. Halverson. *Current Anthropology* 28, 72-3.

BAHN, P.G. 1987b. Excavation of a Palaeolithic plank from Japan. *Nature* 329, p. 110.

BAHN, P.G. 1988. Expecting the Spanish Inquisition: the rejection of Altamira in its 19th century context. Paper for the 5th Int. Conference on Hunting and Gathering Societies, Darwin, September 1988.

BAHN, P.G. 1988a. Where's the beef? The myth of hunting magic in Palaeolithic art. Paper for the 1st AURA (Australian Rock Art Research Association) Congress, Darwin, September 1988.

BAHN, P.G. & BUTLIN, R.K. 1987. Les insectes dans l'art paléolithique: quelques observations nouvelles sur la sauterelle d'Enlène (Ariège). *Actes du Colloque Int. d'Art Mobilier Paléolithique*, Foix/Le Mas d'Azil (in press).

BAHN, P.G. & COLE, G. 1986. La préhistoire pyrénéenne aux Etats-Unis. *Bull. Soc. Préh. Ariège-Pyrénées* 41, 95-149.

BAHN, P.G. & COURAUD, C. 1984. Azilian pebbles - an unsolved mystery. *Endeavour* 8, 156-58.

BAHN, P.G. & OTTE, M. 1985. La poterie 'paléolithique' de Belgique: analyses récentes. *Helinium* 25, 238-41.

de BALBIN BEHRMANN, R. & MOURE ROMANILLO, J.A. 1982. El panel principal de la cueva de Tito Bustillo (Ribadesella, Asturias). *Ars Praehistorica* 1, 47-97.

BANDI, H-G. 1968. Art quaternaire et zoologie, in *Simposio de Arte Rupestre*, Barcelona 1966, 13-19.

BANDI, H-G. 1972. Quelques réflexions sur la nouvelle hypothèse de A. Leroi-Gourhan concernant la signification de l'art quaternaire, in *Santander Symposium*, 309-19.

BANDI, H-G. & DELPORTE, H. 1984. Propulseurs magdaléniens décorés en France et en Suisse. in *Eléments de Pré et Protohistoire européenne, Hommages à J-P. Millotte*, Annales Litt. de l'Univ. de Besançon, 203-17.

BANDI, H-G. *et al* (eds) 1984. *La Contribution de la Zoologie et de l'Ethologie à l'interprétation de l'art des peuples chasseurs préhistoriques.* 3e colloque de la Soc. suisse des Sciences Humaines, Sigriswil 1979. Editions Universitaires: Fribourg.

BARANDIARAN, I. 1968. Rodetes paleolíticos de hueso. *Ampurias* 30, 1-37.

BARANDIARAN, I. 1969. Representaciones de reno en el arte paleolítico español. *Pyrenae* 5, 1-33.

BARANDIARAN, I. 1971. Hueso con grabados paleolíticos en Torre (Oyarzun, Guipúzcoa). *Munibe* 23, 37-69.

BARANDIARAN, I. 1971a. 'Bramaderas' en el Paleolítico superior peninsular. *Pyrenae* 7, 7-18.

BARANDIARAN, I. 1972. Algunas convenciones de representación en las figuras animales del arte paleolítico, in *Santander Symposium* 345-84.

BARANDIARAN, I. 1973. *Arte Mueble del Paleolítico Cantábrico.* Monog. Arq. 14: Zaragoza.

BARANDIARAN, I. 1974. El Glotón (*Gulo gulo L.*) en el arte paleolítico. *Zephyrus* 25, 177-96.

BARANDIARAN, I. 1984. Signos asociados a hocicos de animales en el arte paleolítico. *Veleia* 1, 7-24.

BARRIERE, C. 1976. *L'Art Pariétal de la Grotte de Gargas.* Mémoire III de l'Inst. Art Préhist. Toulouse; British Arch. Reports (Oxford), Int. series No. 14, 2 vols.

BARRIERE, C. & SAHLY, A. 1964. Les empreintes humaines de Lascaux, in *Miscelánea en Homenaje al Abate H. Breuil*, vol. 1, pp. 173-80. Barcelona.

BASLER, D. 1979. Le Paléolithique final en Herzégovine. in *La Fin des Temps Glaciaires en Europe* (D. de Sonneville-Bordes, ed.), vol. 1, 345-55. C.N.R.S.: Paris.

BAULOIS, A. 1980. Les sagaies décorées du Paléolithique supérieur dans la zone franco-cantabrique. *Bull. Soc. Préhist. Ariège* 35, 125-28.

BEASLEY, B. 1986. Les bisons d'argile de la grotte du Tuc d'Audoubert. *Bull. Soc. Préhist. Ariège-Pyrénées* 41, 23-30.

de BEAUNE, S.A. 1984. Comment s'éclairaient les hommes préhistoriques? *La Recherche* 15, No. 152, 247-49.

de BEAUNE, S.A. 1987. *Lampes et Godets au Paléolithique*. XXIIIe Supplément à Gallia Préhistoire.

de BEAUNE, S.A. 1987a. Palaeolithic lamps and their specialization: a hypothesis. *Current Anth.* 28, 569-77.

BEDNARIK, R.G. 1986. Parietal finger markings in Europe and Australia. *Rock Art Research* 3, 30-61 & 159-70.

BEGOUEN, H. 1925/6. Observations nouvelles dans les grottes des Pyrénées, in *Mélanges Gorgianovitch-Kramberger*, 501-9. Zagreb.

BEGOUEN, H. 1929. The magic origin of prehistoric art. *Antiquity* 3, 5-19.

BEGOUEN, H. 1931. La technique des gravures pariétales de quelques grottes pyrénéennes. *XVe Congrès Int. d'Anth. & d'Arch. Préhistorique*, Portugal 1930, 8pp, 13 pl.

BEGOUEN, H. 1932. La Préhistoire à la Société Archéologique du Midi de la France. *Mémoires de la Soc. Arch. du Midi de la France*, 1-11.

BEGOUEN, H. 1942. *De la Lecture des Gravures Préhistoriques. Conseils à mes Etudiants*. Editions du Museum: Toulouse. 10pp.

BEGOUEN, H. 1947. Eloge de M. Emile Cartailhac. *Bull. Soc. Arch. du Midi de la France*, 3e série, 5 (1942-5), 438-51.

BEGOUEN, H. 1952. Séance du 24 juillet. *Bull. Soc. Préh. fr.* 49, 289-91.

BEGOUEN, H. & L. 1934. Quelques plaquettes de pierre gravées ou peintes des cavernes pyrénéennes. *11e Congrès Préhist. de France*, Périgueux, 3pp.

BEGOUEN, H., BEGOUEN, L. & VALLOIS, H. 1937. Une pendeloque faite d'un fragment de mandibule humaine (Epoque magdalénienne). *12e Congrès Préhist. de France*, Toulouse-Foix 1936, 559-64.

BEGOUEN, H. & BREUIL, H. 1958. *Les Cavernes du Volp: Trois-Frères–Tuc d'Audoubert*. Arts et Métiers Graphiques: Paris.

BEGOUEN, H., CASTERET, N. & CAPITAN, L. 1923. La caverne de Montespan (Haute-Garonne). *Revue Anth.* 33, 1-18, 1 pl.

BEGOUEN, J. 1939. De quelques signes gravés et peints des grottes de Montesquieu-Avantès, in *Mélanges Bégouën*, 281-87. Toulouse.

BEGOUEN, L. 1939. Pierres gravées et peintes de l'époque magdalénienne, in *Mélanges Bégouën*, 289-305. Toulouse.

BEGOUEN, R. 1980. La conservation des cavernes du Volp. Son histoire, son bilan. in *Altamira Symposium*, 681-93. Min. de Cultura: Madrid.

BEGOUEN, R. 1984. Les bisons d'argile du Tuc d'Audoubert. *Dossiers de l'Archéologie 87*, octobre, 'Les Premiers Artistes', 77-79.

BEGOUEN, R. & CLOTTES, J. 1979. Le bâton au saumon d'Enlène. *Bull. Soc. Préhist. Ariège* 34, 17-25.

BEGOUEN, R. & CLOTTES, J. 1979a. Galet gravé de la caverne d'Enlène à Montesquieu-Avantès (Ariège). *Caesaraugusta* 49/50, 57-64.

BEGOUEN, R. & CLOTTES, J. 1980. Apports mobiliers dans les cavernes du Volp (Enlène, Les Trois-Frères, Le Tuc d'Audoubert), in *Altamira Symposium*, 157-88. Min. de Cultura: Madrid.

BEGOUEN, R. & CLOTTES, J. 1981. Nouvelles fouilles dans la Salle des Morts de la Caverne d'Enlène à Montesquieu-Avantès (Ariège), in *21e Congrès Préhist. de France*, Montauban/Cahors 1979, vol. 1, 33-56.

BEGOUEN, R. & CLOTTES, J. 1983. El arte mobiliar de las cavernas del Volp (en Montesquieu-Avantès/Ariège). *Revista de Arqueologia* Año IV, No. 27, 6-17.

BEGOUEN, R. & CLOTTES, J. 1984. Un cas d'érotisme préhistorique. *La Recherche* 15, 992-95.

BEGOUEN, R. & CLOTTES, J. 1985. L'art mobilier des Magdaléniens. *Archéologia* 207, novembre, 40-49.

BEGOUEN, R. & CLOTTES, J. 1986/7. Le grand félin des Trois-Frères. *Antiquités Nationales* 18/19, 109-13.

BEGOUEN, R. & CLOTTES, J. 1987. Les Trois-Frères after Breuil. *Antiquity* 61, 180-87.

BEGOUEN, R., CLOTTES, J. & DELPORTE, H. 1977. Le retour du petit bison au Tuc d'Audoubert. *Bull. Soc. Préhist. fr.* 74, 112-20.

BEGOUEN, R. *et al* 1982. Plaquette gravée d'Enlène, Montesquieu-Avantès (Ariège). *Ibid.* 79, 103-12.

BEGOUEN, R. *et al* 1984. Compléments à la grande plaquette gravée d'Enlène. *Ibid.* 81, 142-48.

BEGOUEN, R. *et al* 1984/5. Art mobilier sur support lithique d'Enlène (Montesquieu-Avantès, Ariège), Collection Bégouën du Musée de l'Homme. *Ars Praehistorica* III/IV, 25-80.

BELFER-COHEN, A. & BAR-YOSEF, O. 1981. The Aurignacian at Hayonim Cave. *Paléorient* 7 (2), 19-42.

BELLIER, C. 1981/2. *Contribution à l'étude de l'art paléolithique en Europe occidentale, le contour découpé en os*. Mémoire de licence en histoire de l'art et arch., Univ. libre de Bruxelles, 2 vols.

BELLIER, C. 1984. Contribution à l'étude de l'industrie osseuse préhistorique: les contours découpés du type 'têtes d'herbivores'. *Bull. Soc. royale belge Anth. Préhist.* 95, 21-34.

BELTRAN, A., GAILLI, R. & ROBERT, R. 1973. *La Cueva de Niaux*. Monograf. Arq. No. 16, Zaragoza.

BELTRAN, A., ROBERT, R. & GAILLI, R. 1967. *La Cueva de Bédeilhac*. Monograf. Arq. No. 2, Zaragoza.

BERDIN, M.O. 1970. La répartition des mammouths dans l'art pariétal quaternaire. *Travaux Inst. Art Préhist. Toulouse* 12, 181-367.

BERENGUER ALONSO, M. 1986. Art pariétal paléolithique occidental. Techniques d'expression et identification chronologique. *L'Anthropologie 90*, 665-77.

BIBIKOV, S. 1975. A Stone Age orchestra. *UNESCO Courier*, June, 8-15.

BLANCHARD, J. 1964. Informations recherchées d'après les équidés européens figurés, in *Prehistoric Art of the Western Mediterranean and the Sahara* (L. Pericot & E. Ripoll, eds) 3-34. Viking Fund Publications in Anth. No. 39, New York.

BLANCHARD, J. 1964a. Sélection intentionnelle des belles têtes de cerfs gravées et peintes, in *Miscelánea en Homenaje al abate H. Breuil*, vol. 1, 249-58. Barcelona.

BOHIGAS ROLDAN, R. *et al* 1986. Informe sobre el santuario rupestre paleolítico de la Fuente del Salín (Muñorrodero, Val de San Vicente, Cantabria). *Bol. Cantabro de Espeleología* No. 7, diciembre, 81-98, 1 pl.

BORDES, F. 1952. Sur l'usage probable de la peinture corporelle dans certaines tribus moustériennes. *Bull. Soc. Préhist. française* 49, 169-71.

BORDES, F. 1969. Os percé moustérien et os gravé acheuléen du Pech de l'Azé II. *Quaternaria* 11, 1-6.

BOSINSKI, G. 1973. Le site magdalénien de Gönnersdorf (Commune de Neuwied, Vallée du Rhin Moyen, RFA). *Bull. Soc. Préhist. Ariège* 28, 25-48.

BOSINSKI, G. 1982. *Die Kunst der Eiszeit in Deutschland und in der Schweiz*. Habelt: Bonn.

BOSINSKI, G. 1984. The mammoth engravings of the magdalenian site Gönnersdorf (Rhineland, Germany), in Bandi *et al* 1984, 295-322.

BOSINSKI, G. & FISCHER, G. 1974. *Die Menschendarstellungen von Gönnersdorf der Ausgrabungen von 1968*. Steiner: Wiesbaden.

BOSINSKI, G. & FISCHER, G. 1980. *Mammut- und Pferdedarstellungen von Gönnersdorf*. Steiner: Wiesbaden.

BOUCHUD, J. 1966. *Essai sur le Renne et la Climatologie du Paléolithique Moyen et Supérieur*. Magne: Périgueux.

BOUET, B. *et al* 1986/7. Le Centre d'Information et de Documentation (C.I.D.) H. Breuil. *Antiquités Nationales* 18/19, 9-15.

BOURDELLE, E. 1956. Les parentés morphologiques des Equidés caballins d'après les gravures rupestres du Sud-Ouest de la France. *Mammalia* 20, 22-23.

BOUVIER, J-M. 1976. La Chaire à Calvin, Mouthiers (Charente). Données et problèmes, in *Sud-Ouest (Aquitaine et Charente)*, Livret-Guide de l'Excursion A4, U.I.S.P.P. Nice, 133-36.

BOUVIER, J-M. 1977. *Un Gisement Préhistorique: La Madeleine*. Fanlac: Périgueux.

BOYLE, M. *et al* 1963. Recollections of the Abbé Breuil. *Antiquity* 37, 12-18.

BREUIL, H. 1905. La dégénérescence des figures d'animaux en motifs ornementaux à l'époque du Renne. *C.r. Acad. Inscr. Belles-Lettres*, 105-20.

BREUIL, H. 1909. L'évolution de l'art quaternaire et les travaux d'Edouard Piette. *Revue Arch.* 4e série, 13, 378-411.

BREUIL, H. 1912. L'âge des cavernes et roches ornées de France et d'Espagne. *Revue arch.* 19, 193-234.

BREUIL, H. 1925. Les origines de l'art. *Journal de Psychologie* 22, 289-96.

BREUIL. H. 1926. Les origines de l'art décoratif. *Ibid.* 23, 364-75.

BREUIL, H. 1936/7. De quelques oeuvres d'art magdaléniennes inédites ou peu connues. *I.P.E.K.* 11, 1-16.

BREUIL, H. 1949. Les fresques de la Galerie Vidal à la Caverne de Bédeilhac (Ariège). *Bull. Soc. Préhist. Ariège* 4, 11-16, 11 pl.

BREUIL, H. 1952. *Four Hundred Centuries of Cave Art*. Centre d'Etudes et de Documentation Préhistoriques: Montignac.

BREUIL, H. 1952a. La Caverne de Niaux. Compléments inédits sur sa décoration. *Bull. Soc. Préhist. Ariège* 7, 11-35.

BREUIL, H. 1959. Notre art de l'Epoque du Renne: sa répartition, sa succession, ses musées. in Zervos 1959, pp. 13-20 & 109.

BREUIL, H. 1962. Théories et faits cantabriques relatifs au Paléolithique supérieur et à son art des cavernes. *Munibe* 14, 353-58.

BREUIL, H. 1964. Préface, in *Musée des Antiquités Nationales. Collection Piette.*, by M. Chollot. Musées Nationaux, Paris, pp. 11-13.

BREUIL, H. 1977. La perspective dans les dessins paléolithiques antérieurs au Solutréen, in *Cougnac, grotte peinte*, by L. Méroc & J. Mazet, 53-61. Editions des grottes de Cougnac: Gourdon.

BREUIL, H. & BEGOUEN, H. 1937. Quelques oiseaux inédits ou méconnus de l'art préhistorique. *12e Congrès Préhist. de France*, Toulouse-Foix 1936, 475-88.

BREUIL, H. & LANTIER, R. 1959. *Les Hommes de la Pierre Ancienne*. Payot: Paris. 2nd edition.

BREUIL, H., NOUGIER, L-R & ROBERT, R. 1956. Le 'lissoir aux ours' de la grotte de la Vache, à Alliat, et l'ours dans l'art franco-cantabrique occidental. *Bull. Soc. Préhist. Ariège* 11, 15-78, 7 pl. (& 12, 1957, 53-54).

BREUIL, H. & OBERMAIER, H. 1935. *The Cave of Altamira at Santillana del Mar*, Spain. Tipografía de Archivos: Madrid.

BREUIL, H. & de SAINT PERIER, R. 1927. *Les Poissons, les Batraciens et les Reptiles dans l'Art Quaternaire*. Archives de l'Inst. Paléont. Humaine 2, Paris.

BREWSTER, R. 1986. Aboriginal cave paintings among oldest in world. *Popular Archaeology* 7 (6), p. 44.

BRIELLE, G. 1968. Les bouquetins dans l'art pariétal quaternaire pyrénéen. *Travaux Inst. Art Préhist. Toulouse* 10, 31-96.

BRODRICK, A.H. 1963. *The Abbé Breuil, Prehistorian; a Biography*. Hutchinson: London.

BROWN, S. 1987. The 1986 AURA Franco-Cantabrian field trip: summary report. *Rock Art Research* 4 (1), 56-60.

BRUNET, J. 1981. Niaux: Du charbon de bois dans les peintures. *Archéologia* 156, juillet, p. 72.

BRUNET, J. 1982. La Grotte de Niaux. *Archéologia* 162, janvier, p. 74.

BUISSON, D. & PINÇON, G. 1986/7. Nouvelle lecture d'un galet gravé de Gourdan et essai d'analyse des figurations d'oiseaux dans l'art paléolithique français. *Antiquités Nationales* 18/19, 75-90.

BUTZER, K.W. *et al* 1979. Dating and context of rock engravings in southern Africa. *Science* 203, 1201-14.

CABRE AGUILO, J. 1934. *Las Cuevas de Los Casares y de la Hoz*. Archivo Esp. de Arte y Arq. 30, Madrid.

CAMPBELL, J.B. 1977. *The Upper Palaeolithic of Britain*. 2 vols, Clarendon Press: Oxford.

CAPDEVILLE, E. 1986. Aperçus sur le problème des signes tectiformes dans l'art pariétal paléolithique supérieur d'Europe. *Travaux de l'Institut d'Art Préhistorique de Toulouse* 28, 59-104.

CAPITAN, L. & BOUYSSONIE, J. 1924. *Un atelier d'art préhistorique. Limeuil. Son gisement à gravures sur pierres de l'âge du Renne*. Nourry: Paris.

CAPITAN, L., BREUIL, H. & PEYRONY, D. 1910. *La Caverne de Font-de-Gaume aux Eyzies (Dordogne)*. Monaco.

CAPITAN, L., BREUIL, H. & PEYRONY, D. 1924. *Les Combarelles aux Eyzies (Dordogne)*. Masson: Paris.

CAPITAN, L. & PEYRONY, D. 1928. *La Madeleine. Son gisement, son industrie, ses oeuvres d'art*. Nourry: Paris.

CARALP, E., NOUGIER, L-R. & ROBERT, R. 1973. Le thème du 'mammifère aux poissons' dans l'art magdalénien. *Bull. Soc. Préhist. Ariège* 28, 11-23.

CARCAUZON, C. 1984. Une nouvelle découverte en Dordogne: la grotte préhistorique de Fronsac. *Revue Archéologique Sites* 22, août, 7-15.

CARCAUZON, C. 1986. La grotte de Font-Bargeix. *Bull. Soc. Hist. et Arch. du Périgord* 113, 191-98.

CARCAUZON, C. 1988. Découverte de quatre grottes ornées en Périgord. *Archéologia* 235, 16-24.

CARCIUMARU, M. & BITIRI, M. 1983. Peintures rupestres de la grotte Cuciulat (Roumanie). *Bull. Soc. Préhist. française* 80, 94-96.

CARTAILHAC, E. 1902. Les cavernes ornées de dessins. La grotte d'Altamira, Espagne. 'Mea culpa d'un sceptique'. *L'Anthropologie* 13, 348-54.

CARTAILHAC, E. 1908. Les plus anciens artistes de l'Humanité. *Revue des Pyrénées* 20, 509-27.

CARTAILHAC, E. & BREUIL, H. 1905. Les peintures et gravures murales des cavernes pyrénéennes: Marsoulas. *L'Anth.* 16, 431-43.

CARTAILHAC, E. & BREUIL, H. 1906. *La Caverne d'Altamira à Santillane, près Santander (Espagne)*. Imp. de Monaco.

CASADO LOPEZ, M.P. 1977. *Los Signos en el Arte Paleolítico de la Península Ibérica*. Monografías Arq. 20, Zaragoza.

CATTELAIN, P. 1977/8. *Les propulseurs au Paléolithique Supérieur. Essai d'un inventaire descriptif et critique*. Mémoire de licence en histoire de l'art et arch., Univ. libre de Bruxelles, 2 vols.

CHABREDIER, L. 1975. Les gravures paléolithiques de la grotte d'Ebbou (Ardèche). *Archéocivilisation*, nlle série, numéro spécial, 14/15.

CHALOUPKA, G. 1984. *From Palaeoart to Casual Paintings*. Monograph 1, Northern Territory Museum of Arts and Sciences: Darwin.

CHOLLOT-LEGOUX, M. 1961. Une figuration magdalénienne de dépouille animale. *Antiquités Nationales et Internationales* IIe année, mars–juin, p. 13 & 1 pl.

CHOLLOT-LEGOUX, M. 1964. *Musée des Antiquités Nationales: Collection Piette (Art Mobilier Préhistorique)*. Musées Nationaux: Paris.

CHOLLOT-VARAGNAC, M. 1980. *Les Origines du Graphisme Symbolique*. Singer-Polignac: Paris.

CLARK, L.H. 1966. *They Sang For Horses: The Impact of the Horse on Navajo and Apache Folklore*. Univ. of Arizona Press.

CLEGG, J. 1985. Ethnography, pictures and theory. Paper for A.A.A. Conference, Valla, 20pp.

CLEGG, J. 1986. Berkeley Addendum to the above, 7pp. See also 'Prehistoric messages in art and the Pink Panther', in *Univ. of Sydney News* 29/4/86, p. 89.

CLOTTES, J. 1986. La détermination des figurations humaines et animales dans l'art paléolithique européen, in *Cultural Attitudes to Animals including Birds, Fish and Invertebrates*, vol. 3, 39pp. World Arch. Congress, Allen & Unwin: London.

COLLINS, D. & ONIANS, J. 1978. The origins of art. *Art History* 1, 1-25.

COLLOT, F. *et al* La théorie de l'information en Préhistoire. *Bio-Mathématique* 77, 32pp.

COMBIER, J. 1984. Grottes ornées de l'Ardèche, in *Les Premiers Artistes*, Dossier de l'Arch. 87, 80-86.

COMBIER, J., DROUOT, E. & HUCHARD, P. 1958. Les grottes solutréennes à gravures pariétales du canyon inférieur de l'Ardèche. *Mém. Soc. Préhist. fr.* 5, 61-117.

CONKEY, M.W. 1978. Style and information in cultural evolution: toward a predictive model for the Paleolithic, in *Social Archaeology, Beyond Subsistence and Dating* (C.L. Redman *et al*, eds), 61-85. Academic Press: New York/London.

CONKEY, M.W. 1980. The identification of prehistoric hunter-gatherer aggregation sites; the case of Altamira. *Current Anth.* 21, 609-30.

CONKEY, M.W. 1981. A century of Palaeolithic cave art. *Archaeology* 34, 20-28

CONKEY, M.W. 1983. On the origins of Paleolithic art: a review and some critical thoughts, in *The Mousterian Legacy* (E. Trinkaus, ed.), 201-27. British Arch. Reports, Int series 164: Oxford.

CONKEY, M.W. 1984. To find ourselves: art and social geography of prehistoric hunter-gatherers, in *Past and Present in Hunter-Gatherer Studies* (C. Schrire, ed.), 253-76. Academic Press: New York/London.

CONKEY, M.W. 1985. Ritual communication, social elaboration and the variable trajectories of Paleolithic material culture, in *Prehistoric Hunter-Gatherers: the Emergence of Social and Cultural Complexity* (T.D. Price & J.A. Brown, eds), 299-323. Academic Press: New York/London.

COURAUD, C. 1982. Techniques de peintures préhistoriques: Expériences. *Information Couleur* 19, 3-6.

COURAUD, C. 1983. Pour une étude méthodologique des colorants préhistoriques. *Bull. Soc. Préhist. fr.* 80, 104-10.

COURAUD, C. 1983a. La couleur dans l'art paléolithique. *Information Couleur* 22, 5-9.

COURAUD, C. 1985. *L'Art Azilien. Origine - Survivance*. XXe Supplément à Gallia Préhistoire.

DAMS, L. 1978. *L'Art Paléolithique de la Caverne de la Pileta*. Akademische Druck- u. Verlagsanstalt: Graz.

DAMS, L. 1984. Preliminary findings at the 'organ' sanctuary in the cave of Nerja, Málaga, Spain. *Oxford Journal of Arch.* 3, 1-14.

DAMS, L. 1985. Palaeolithic lithophones: descriptions and comparisons. *Oxford Journal of Arch.* 4, 31-46.

DAMS, L. 1987. Poissons et contours de type pisciforme dans l'art pariétal paléolithique. *Bull. Soc. Royale belge Anth. Préhist.* 98, 81-132. (Also 'Fish images in palaeolithic cave art' in *Archaeology Today* 8 (2), 16-20).

DAMS, L. 1987a. *L'Art Paléolithique de la Grotte de Nerja (Málaga, Espagne)*. British Arch. Reports, Int. Series No. 385, Oxford.

DASTUGUE, J. & DE LUMLEY, M.A. 1976. Les maladies des hommes préhistoriques du Paléolithique supérieur et du Mésolithique, in *La Préhistoire française*, vol. I:1, 612-22 (H. de Lumley, ed.). C.N.R.S.: Paris.

DAUVOIS, M. 1977. Travail expérimental de l'ivoire: sculpture d'une statuette féminine, in *Méthodologie appliquée à l'industrie de l'os préhistorique* (H. Camps-Fabrer, ed.), 270-73. C.N.R.S.: Paris.

DAVENPORT, D. & JOCHIM, M.A. 1988. A note on the scene in the shaft at Lascaux. *Antiquity* 62 (in press).

DAVIS, W. 1986. The origins of image making. *Current Anthropology* 27, 193-215, 371, 515-16.

DAVIS, W. 1987. Replication and depiction in Paleolithic art. *Representations* 19, 111-47.

DECOUVERTES DE L'ART DES GROTTES ET DES ABRIS. 1984. Catalogue de l'Exposition, 59pp. Réjou: Périgueux.

DELCOURT-VLAEMINCK, M. 1975. Les représentations végétales dans l'art du Paléolithique supérieur. *Paléontologie et Préhistoire: Rev. Soc. Tournaisienne de Géol., Préh. et Arch.* 31, 70-96.

DELLUC, B. & G. 1973. Quelques figurations paléolithiques inédites des environs des Eyzies (Dordogne): Grottes Archambeau, du Roc et de la Mouthe. *Gallia Préhistoire* 16, 201-9.

DELLUC, B. & G. 1978. Les manifestations graphiques aurignaciens sur support rocheux des environs des Eyzies (Dordogne). *Gallia Préhistoire* 21, 213-438.

DELLUC, B. & G. 1981. Une visite à la grotte de Rouffignac en 1759. *Bull. Soc. Hist. et Arch. du Périgord* 108, 3-11.

DELLUC, B. & G. 1981a. La grotte ornée de Comarque à Sireuil (Dordogne). *Gallia Préhistoire* 24, 1-97.

DELLUC. B. & G. 1981b. Les signes en 'empreinte' du début du Paléolithique supérieur, in *21e Congrès Préhist. de France*, vol. 2, 111-16.

DELLUC, B. & G. 1983. Les grottes ornées de Domme (Dordogne): La Martine, Le Mammouth et le Pigeonnier. *Gallia Préhistoire* 26, 7-80.

DELLUC, B. & G. 1983a. La Croze à Gontran, grotte ornée aux Eyzies-de-Tayac (Dordogne). *Ars Praehistorica* 2, 13-48.

DELLUC, B. & G. 1984. L'art pariétal avant Lascaux. *Dossiers de l'Archéologie* 87, octobre, 'Les Premiers Artistes', 52-60.

DELLUC, B. & G. 1984a. Lascaux II, a faithful copy. *Antiquity* 58, 194-96, 2 pl.

DELLUC, B. & G. 1984b. Lecture analytique des supports rocheux gravés et relevé synthétique. *L'Anthropologie* 88, 519-29.

DELLUC, B. & G. 1984c. Faune figurée et faune consommée: une magie de la chasse? *Dossiers de l'Arch.* 87, oct., 'Les Premiers Artistes', 28-29.

DELLUC, B. & G. 1985. De l'empreinte au signe, in *Traces et Messages de la Préhistoire*, Dossiers de l'Arch. 90, 56-62.

DELPORTE, H. 1962. Le problème des statuettes féminines dans le leptolithique occidental. *Mitteilungen der Anth. Gesellschaft in Wien* 92, 53-60.

DELPORTE, H. 1971. A propos du style des figurations féminines gravettiennes. *Antiquités Nationales* 3, 5-20.

DELPORTE, H. 1973. Les techniques de la gravure paléolithique, in *Estudios dedicados al Prof. Dr. Luis Pericot*, 119-29. Inst. de Arq. y Preh., Univ. de Barcelona.

DELPORTE, H. 1979. *L'Image de la Femme dans l'Art Préhistorique*. Picard: Paris.

DELPORTE, H. 1984. L'art mobilier et ses rapports avec la faune paléolithique, in Bandi *et al* 1984, 111-42.

DELPORTE, H. 1987. *Piette, pionnier de la Préhistoire*. Picard: Paris.

DELPORTE, H., KANDEL, D. & PINÇON, G. 1986. Le C.I.D. Breuil, domaine A.P.M.: Un système documentaire sur l'Art Paléolithique Mobilier. *Bull. Soc. Préhist. française* 83, 299-303.

DELPORTE, H. & MONS, L. 1975. Omoplate décorée du Mas d'Azil (Ariège). *Antiquités Nationales* 7, 14-23.

DELPORTE, H. & MONS, L. 1980. Gravure sur os et argile, in *Revivre la Préhistoire*, Dossier de l'Archéologie 46, 40-45.

DESBROSSE, R., FERRIER, J. & TABORIN, Y. 1976. La parure. in *La Préhistoire française*, I:1 (H. de Lumley, ed.), 710-13. C.N.R.S.: Paris.

DEWEZ, M. 1974. New hypotheses concerning two engraved bones from La Grotte de Remouchamps, Belgium. *World Arch.* 5, 337-45, 1 pl.

DJINDJIAN, F. & PINÇON, G. 1986. Un exemple de banque de données sur microserveur Vidéotex: L'Art Pariétal Paléolithique. *Bull. Soc. Préhist. française* 83, 332-34.

DORN, R.I., NOBBS, M. & CAHILL, T.A. (in press). Cation-ratio dating of rock engravings from the Olary Province of arid South Australia. *Antiquity* (forthcoming).

DORTCH, C. 1979. Australia's oldest known ornaments. *Antiquity* 53, 39-43.

DORTCH, C. 1984. *Devil's Lair: A study in prehistory*. Western Australia Museum: Perth.

DUHARD, J-P. 1987. Le soi-disant 'hermaphrodite' de Grimaldi: une nouvelle interprétation obstétricale. *Bull. Soc. Anth. Sud-Ouest* 22, 139-44.

DUHARD, J-P. 1987a. La Statuette de Monpazier représente-elle une parturiente? *Bull. Soc. Préhist. Ariège-Pyrénées* 42, 155-63.

DUHARD, J-P. & ROUSSOT, A. 1988. Le gland pénien sculpté de Laussel (Dordogne). *Bull. Soc. Préhist. française* 85, 41-44.

EASTHAM, A. 1986. The season or the symbol: the evidence of swallows in the Paleolithic of Western Europe. Paper for International Council for Archaeozoology Conference, Bordeaux 1986, 14pp.

EASTHAM, A. & M. 1979. The wall art of the Franco-Cantabrian deep caves. *Art History* 2, 365-87.

EATON, R.L. 1978. The evolution of trophy hunting. *Carnivore* 1, 110-21.

EPPEL, F. 1972. Les objets d'art paléolithique en Autriche. *Bull. Soc. Préhist. Ariège* 27, 73-81.

d'ERRICO, F. 1987. Nouveaux indices et nouvelles techniques microscopiques pour la lecture de l'art gravé mobilier. *C.R. Acad. Sc. Paris* 304, Série II, No. 13, 761-64, 1 pl.

FAGES, G. & MOURER-CHAUVIRE, C. 1983. La flûte en os d'oiseau de la grotte sépulcrale de Veyreau (Aveyron) et inventaire des flûtes préhistoriques d'Europe. in *La Faune et l'Homme Préhistoriques*, Mém. 16 de la Soc. Préhist. française, 95-103.

FAURE, M. 1978. Révision critique d'une collection de gravures mobilières paléolithiques: les galets et les os gravés de la Colombière (Neuville-sur-Ain, Ain, France). *Nouv. Arch. Mus. Hist. nat. Lyon* fasc. 16, 41-99, 6 pl.

FORBES, A. & CROWDER, T.R. 1979. The problem of Franco-Cantabrian abstract signs: agenda for a new approach. *World Arch.* 10, 350-66, 2 pl.

FORTEA, F.J. 1978. Arte paleolítico del Mediterraneo español. *Trabajos de Prehistoria* 35, 99-149.

FORTEA, J. 1981. Investigaciones en la cuenca media del Nalón, Asturias (España). Noticia y primeros resultados. *Zephyrus* 32/33, 5-16.

FREEMAN, L.G. 1978. Mamut, jabalí y bisonte en Altamira: reinterpretaciones sugeridas por la historia natural, in *Curso de Arte Rupestre Paleolítico* (A. Beltrán, ed.), 157-79. Universidad Int. 'Menéndez Pelayo': Santander.

FREEMAN, L.G. 1984. Techniques of figure enhancement in Paleolithic cave art, in *Scripta Praehistorica. Francisco Jordá Oblata*, 209-32. Univ. de Salamanca.

FREEMAN, L.G. & GONZALEZ ECHEGARAY, J. 1983. Tally-marked bone from Mousterian levels at Cueva Morín (Santander, Spain), in *Homenaje al Prof. M. Almagro Basch*, I, 143-47. Min. de Cultura: Madrid.

FREUND, G. 1957. L'art aurignacien en Europe centrale. *Bull. Soc. Préhist. Ariège* 12, 55-78, 2 pl.

FROLOV, B.A. 1977/9. Numbers in Paleolithic graphic art and the initial stages in the development of mathematics. *Soviet Anth. and Arch.* 16, 1977/8, 142-66; and 17, 1978/9, 73-93, 41-74 & 61-113.

FROLOV, B.A. 1979. Les bases cognitives de l'art paléolithique, in *Valcamonica Symposium III. The Intellectual Expressions of Prehistoric Man: Art and Religion*, 295-98.

FROLOV, B.A. 1981. L'art paléolithique: Préhistoire de la science? in *Arte Paleolítico*, Comisión XI, Xth Congress UISPP, Mexico City, 60-81.

GAMBLE, C. 1982. Interaction and alliance in Palaeolithic society. *Man* 17, 92-107.

GARCIA, M. 1979. Les silicones élastomères R.T.V. appliqués aux relevés de vestiges préhistoriques (Arts, Empreintes Humaines et Animales). *L'Anthropologie* 83, 5-42 & 189-222.

GARCIA GUINEA, M.A. 1979. *Altamira y otras cuevas de Cantabria*. Silex: Madrid.

GAUSSEN, J. 1964. *La Grotte Ornée de Gabillou (près Mussidan, Dordogne)*. Mémoire 4, Inst. de Préhist. de l'Univ. de Bordeaux.

GEOFFROY, C. 1974. La couleur dans l'art pariétal paléolithique. *Cahiers du Centre de Recherches Préhist.* 3, 45-64.

GIEDION, S. 1965. *L'Eternel Présent. La Naissance de l'Art*. Editions de la Connaissance: Bruxelles.

GIMBUTAS, M. 1981. Vulvas, breasts and buttocks of the Goddess Creatress: Commentary on the origins of art, in *The Shape of the Past: Studies in Honor of Franklin D. Murphy* (G. Buccellati & C. Speroni, eds), 15-42. Inst. of Arch.: Los Angeles.

GLADKIH, M.I., KORNIETZ, N.L. & SOFFER, O. 1984. Mammoth-bone dwellings on the Russian Plain. *Scientific American* 251, 136-42.

GLORY, A. 1964. La stratigraphie des peintures à Lascaux, in *Miscelánea en Homenaje al Abate H. Breuil*, vol. 1, 449-55. Barcelona.

GLORY, A. 1968. L'énigme de l'art quaternaire peut-elle être résolue par la théorie du culte des ongones? in *Simposio de Arte Rupestre*, Barcelona 1966, 25-60.

GONZALEZ ECHEGARAY, J. 1968. Sobre la datación de los santuarios paleolíticos, in *Simposio de Arte Rupestre*, Barcelona 1966, 61-65.

GONZALEZ ECHEGARAY, J. 1972. Notas para el estudio cronológico del arte rupestre de la cueva del Castillo, in *Santander Symposium* 409-22.

GONZALEZ ECHEGARAY, J. 1974. *Pinturas y grabados de la cueva de las Chimeneas (Puente Viesgo, Santander)*. Monografías de Arte Rupestre, Arte Paleolítico No. 2, Barcelona.

GONZALEZ GARCIA, R. 1985. *Aproximació al Desenvolupament i Situació de les Manifestacions Artístiques Quaternàries: a les cavitats del Monte del Castillo*. Tesis de Llicenciatura, Univ. de Barcelona, Fac. de Geografia i Historia, Dept. d'Historia de l'Art. 547pp.

GONZALEZ GARCIA, R. 1987. Organisation, distribution and typology of the cave art of Monte del Castillo, Spain. *Rock Art Research* 4, 127-36.

GOREN-INBAR, N. 1986. A figurine from the Acheulian site of Berekhat Ram. *Mi'Tekufat Ha'Even* 19, 7-12.

GRAINDOR, M.J. & MARTIN, Y. 1972. *L'Art Préhistorique de Gouy*. Presses de la Cité: Paris.

GRAZIOSI, P. 1960. *Palaeolithic Art*. Faber: London.

GRAZIOSI, P. 1973. *L'Arte Preistorica in Italia*. Sanzoni: Florence.

GUIDON, N. & DELIBRIAS, G. 1986. Carbon-14 dates point to man in the Americas 32,000 years ago. *Nature* 321, 769-71.

GUTHRIE, R.D. 1984. Ethological observations from Palaeolithic art, in Bandi *et al* 1984, 35-74.

HAHN, J. 1971. La statuette masculine de la grotte du Hohlenstein-Stadel (Wurtemberg). *L'Anthropologie* 75, 233-44.

HAHN, J. 1972. Aurignacian signs, pendants and art objects in central and eastern Europe. *World Arch.* 3, 252-66, 2 pl.

HAHN, J. 1979. Elfenbeinplastiken des Aurignacien aus dem Geissenklösterle. *Archäologisches Korrespondenzblatt* 9, 135-42, 2 pl.

HAHN, J. 1982. Demi-relief aurignacien en ivoire de la grotte Geissenklösterle, près d'Ulm (Allemagne fédérale). *Bull. Soc. préhist. fr.* 79, 73-77.

HAHN, J. 1984. L'art mobilier aurignacien en Allemagne du Sud-Ouest: Essai d'analyse zoologique et éthologique, in Bandi *et al* 1984, 283-93.

HALLAM, S.J. 1971. Roof marking in the 'Orchestra Shell' Cave, Wanneroo, near Perth, Western Australia. *Mankind* 8, 90-103.

HALVERSON, J. 1987. Art for art's sake in the Paleolithic. *Current Anth.* 28, 63-89, 203-05.

HAMMOND, N. 1974. Paleolithic mammalian faunas and parietal art in Cantabria: a comment on Freeman. *American Antiquity* 39, 618-19.

HARLE, E. 1881. La grotte d'Altamira, près de Santander (Espagne). *Matériaux pour l'Histoire Primitive et Naturelle de l'Homme* 2e série, XII, 275-83, 1 pl.

HARRISON, R.A. 1978. A pierced reindeer phalanx from Banwell Bone Cave and some experimental work on phalangeal whistles. *Proc. Univ. Bristol Spel. Soc.* 15, 7-22, 1 pl.

HENNIG, H. 1960. L'art magdalénien en Europe centrale. *Bull. Soc. Préh. Ariège* 15, 24-45.

HORNBLOWER, G.D. 1951. The origin of pictorial art. *Man* 51, 2-3.

JANSSENS, P.A. 1957. Medical views on prehistoric representations of human hands. *Medical History* 1, 318-22, 2 pl.

JELINEK, J. 1975. *Encyclopédie illustrée de l'Homme Préhistorique*. Gründ: Paris.

JIA LANPO 1980. *Early Man in China*. Foreign Languages Press: Beijing.

JOCHIM, M. 1983. Palaeolithic cave art in ecological perspective, in *Hunter-Gatherer Economy in Prehistory* (G.N. Bailey, ed.), 212-19. Cambridge Univ. Press.

JORDA CERDA, F. 1964. El arte rupestre paleolítico de la región cantábrica: nueva secuencia cronológico-cultural, in *Prehistoric Art of the Western Mediterranean and the Sahara* (L. Pericot & E. Ripoll, eds), 47-81. Viking Fund Publications in Anth., No. 39, New York.

JORDA CERDA, F. 1964a. Sobre técnicas, temas y etapas del Arte Paleolítico de la Región Cantábrica. *Zephyrus* 15, 5-25.

JORDA CERDA, F. 1972. Las superposiciones en el gran techo de Altamira, in *Santander Symposium*, 423-56.

JORDA CERDA, F. 1979. Sur des sanctuaires monothématiques dans l'art rupestre cantabrique, in *Valcamonica Symposium III. The Intellectual Expressions of Prehistoric Man: Art and Religion*, 331-48.

JORDA CERDA, F. 1980. El gran techo de Altamira y sus santuarios superpuestos, in *Altamira Symposium*, 277-87. Min. de Cultura: Madrid.

JORDA CERDA, F. 1983. El mamut en el arte paleolítico peninsular y la hierogamia de Los Casares, in *Homenaje al Prof. M. Almagro Basch*, vol. 1, 265-72. Min. de Cultura: Madrid.

JORDA CERDA, F. 1985. Los grabados de Mazouco, los santuarios monotemáticos y los animales dominantes en el arte paleolítico peninsular. *Revista de Guimarães* 94, 23pp.

JORDA CERDA, F. 1987. Sobre figuras rupestres paleolíticas de posibles caballos domesticados. *Archivo de Prehistoria Levantina* 17, 49-58.

JORGE, S.O. *et al* 1981. Gravuras rupestres de Mazouco (Freixo de Espada à Cinta). *Arqueología* (Porto) 3, 3-12.

JORGE, S.O. *et al* 1982. Descoberta de gravuras rupestres em Mazouco, Freixo de Espada-à-Cinta (Portugal). *Zephyrus* 34/35, 65-70.

JUDSON, S. 1959. Paleolithic paint. *Science* 130, p. 708.

JULIEN, M. & LAVALLEE, D. 1987. Les chasseurs de la préhistoire, in *Ancien Pérou: Vie, Pouvoir et Mort*. Exhibition catalogue, Musée de l'Homme, pp. 45-60. Nathan: Paris.

KEHOE, T.F. 1987. Corralling life. *Wisconsin Academy Review*, March, 45-48.

KEHOE, T.F. (in press) Corralling life. *Plains Anthropologist*

KLIMA, B. 1982. Neue Forschungsergebnisse in Dolní Věstonice, in *Aurignacien et Gravettien en Europe, Fascicule II, Cracovie/Nitra 1980*. Etudes et Recherches Arch. de l'Univ. de Liège, 13, 171-80.

KLIMA, B. 1983. Une nouvelle statuette paléolithique à Dolní Věstonice. *Bull. Soc. Préhist. fr.* 80, 176-78.

KLIMA, B. 1984. Les représentations animales du Paléolithique Supérieur de Dolní Věstonice (Tchécoslovaquie), in Bandi *et al* 1984, 323-32.

KOBY, F.E. 1954. Y a-t-il eu, à Lascaux, 'un Bos longifrons'? *Bull. Soc. Préhist. fr.* 51, 434-41.

KOBY, F.E. 1968. Les 'rennes' de Tursac paraissent être plutôt des daims. *Bull. Soc. Préhist. Ariège* 23, 123-30.

KOZLOWSKI, J.K. (ed.) 1982. *Excavation in the Bacho Kiro Cave (Bulgaria): Final Report*. Panstwowe Wydawnictwo Naukowe: Warsaw.

KRAFT, J.C. & THOMAS, R.A. 1976. Early man at Holly Oak, Delaware. *Science* 192, 756-61.

KÜHN, H. 1955. *On the Track of Prehistoric Man*. Hutchinson: London.

KÜHN, H. 1971. *Die Felsbilder Europas*. Kohlhammer: Stuttgart.

KUMAR, G., NARVARE, G. & PANCHOLI, R. 1988. Engraved ostrich eggshell objects: new evidence of Upper Palaeolithic art in India. *Rock Art Research* 5 (1), 43-53.

KURTEN, B. 1986. *How to Deep-Freeze a Mammoth.* Columbia Univ. Press: New York.

LAMING-EMPERAIRE, A. 1959. *Lascaux. Paintings and Engravings.* Pelican: Harmondsworth.
LAMING-EMPERAIRE, A. 1962. *La Signification de l'Art Rupestre Paléolithique.* Picard: Paris.
LAMING-EMPERAIRE, A. 1970. Système de penser et organisation sociale dans l'art rupestre paléolithique, in *L'Homme de Cro-Magnon,* 197-211. Arts et Métiers Graphiques: Paris.
LAMING-EMPERAIRE, A. 1971. Une hypothèse de travail pour une nouvelle approche des sociétés préhistoriques, in *Mélanges Varagnac,* 541-51. Ecole Pratique des Hautes Etudes: Paris.
LAMING-EMPERAIRE, A. 1972. Art rupestre et organisation sociale, in *Santander Symposium,* 65-82.
LARTET, E. 1861. Nouvelles recherches sur la coexistence de l'Homme et des grands mammifères fossiles. *Annales des Sciences Naturelles* (Zoologie), 4e série, XV, 177-253, 3 pl.
LARTET, E. & CHRISTY, H. 1864. Sur des figures d'animaux gravées ou sculptées et autres produits d'art et d'industrie rapportables aux temps primordiaux de la période humaine. *Revue Arch.* IX, 233-67, 2 pl.
LARTET, E. & CHRISTY, H. 1875. *Reliquiae Aquitanicae.* Jones: London. (1865-1875).
LAURENT, P. 1963. La tête humaine gravée sur bois de renne de la grotte du Placard (Charente). *L'Anthropologie* 67, 563-69.
LAURENT, P. 1971. Iconographie et copies successives: la gravure anthropomorphe du Placard. *Mém. Soc. Arch. Hist. Charente,* 215-28.
LEASON, P.A. 1939. A new view of the Western European group of Quaternary cave art. *Proc. Prehist. Society* 5, 51-60.
LEASON, P.A. 1956. Obvious facts of Quaternary cave art. *Medical and Biological Illustration* 6 (4), 209-14.
LEGUAY, L. 1877. Les procédés employés pour la gravure et la sculpture des os avec le silex. *Bull. Soc. Anth. Paris* 2e série, 12, 280-96.
LEGUAY, L. 1882. Sur la gravure des os au moyen du silex. *C.r. Assoc. française pour l'Avancement des Sciences* 11, La Rochelle, 677-79.
LEJEUNE, M. 1981. *L'Utilisation des Accidents Naturels dans le tracé des figurations pariétales du Paléolithique Supérieur franco-cantabrique.* Mémoire de fin d'études, Univ. de Liège, 274pp.
LEJEUNE, M. 1987. *L'Art Mobilier Paléolithique et Mésolithique en Belgique.* Artefacts 4, Editions du Centre d'Etudes et de Documentation Archéologiques: Treignes-Viroinval, Belgium.

LEJEUNE, M. (in press) *L'Art Pariétal dans son Contexte Naturel.* Etudes et Recherches Arch. de l'Univ. de Liège 14.
LE MORT, F. 1985. Un exemple de modification intentionnelle: la dent humaine perforée de Saint-Germain-la-Rivière (Pal. Sup.). *Bull. Soc. Préhist. fr.* 82, 190-92.
LEONARDI, P. 1983. Incisioni musteriane del Riparo Tagliente in Valpantena nei Monti Lessini presso Verona (Italia), in *Homenaje al Prof. M. Almagro Basch,* vol. 1, 149-54. Min. de Cultura: Madrid.
LEROI-GOURHAN, And. 1958. La fonction des signes dans les sanctuaires paléolithiques. *Bull. Soc. Préhist. fr.* 55, 307-21.
LEROI-GOURHAN, And. 1958a. Le symbolisme des grands signes dans l'art pariétal paléolithique. *Ibid.* 55, 384-98.
LEROI-GOURHAN, And. 1958b. Répartition et groupement des animaux dans l'art pariétal paléolithique. *Ibid.* 55, 515-28.
LEROI-GOURHAN, And. 1962. Chronologie de l'art paléolithique, in *Atti del VI Congresso UISPP,* Rome, vol. 3, 341-45.
LEROI-GOURHAN, And. 1966. Réflexions de méthode sur l'art paléolithique. *Bull. Soc. Préhist. fr.* 63, 35-49.
LEROI-GOURHAN, And. 1967. Les mains de Gargas. Essai pour une étude d'ensemble. *Ibid.* 64, 107-22.
LEROI-GOURHAN, And. 1968. Les signes pariétaux du Paléolithique supérieur franco-cantabrique, in *Simposio de Arte Rupestre,* Barcelona 1966, 67-77.
LEROI-GOURHAN, And. 1971. *Préhistoire de l'Art Occidental.* Mazenod: Paris. 2e Edition.
LEROI-GOURHAN, And. 1972. Considérations sur l'organisation spatiale des figures animales dans l'art pariétal paléolithique, in *Santander Symposium* 281-308.
LEROI-GOURHAN, And. 1978. Le cheval sur galet de la galerie Breuil au Mas d'Azil (Ariège). *Gallia Préh.* 21, 439-45.
LEROI-GOURHAN, And. 1980. Les signes pariétaux comme 'marqueurs' ethniques, in *Altamira Symposium* 289-94. Min. de Cultura: Madrid.
LEROI-GOURHAN, And. 1981. Le cas de Lascaux. *Monuments Historiques* 118, 24-32.
LEROI-GOURHAN, And. 1982. *The Dawn of European Art.* Cambridge Univ. Press.
LEROI-GOURHAN, And. 1983. Les entités imaginaires. Esquisse d'une recherche sur les monstres pariétaux paléolithiques, in *Homenaje al Prof. M. Almagro Basch,* vol. 1, 251-63. Min. de Cultura: Madrid.

LEROI-GOURHAN, And. 1984. Le réalisme du comportement dans l'art paléolithique d'Europe de l'Ouest, in Bandi *et al* 1984, 75-90.

LEROI-GOURHAN, And. 1984a. *Simbolos, Artes y Creencias de la Prehistoria.* Istmo: Madrid.

LEROI-GOURHAN, And. 1984b. *Arte y Grafismo en la Europa Prehistórica.* Istmo: Madrid.

LEROI-GOURHAN, Arl. 1983. Du fond des grottes aux terrasses ensoleillées, in *Homenaje al Prof. M. Almagro Basch*, vol. 1, 239-49. Min. de Cultura: Madrid.

LEROI-GOURHAN, Arl. & ALLAIN, J. (eds). 1979. *Lascaux Inconnu.* XIIe Suppl. à Gallia Préhistoire.

LEROY-PROST, C. 1984. L'art mobilier, in *Les Premiers Artistes*, Dossier de l'Archéologie 87, 45-51.

LEVEQUE, F. & VANDERMEERSCH, B. 1980. Découverte de restes humains dans un niveau castelperronien à Saint-Césaire (Charente-Maritime). *C.r. Académie des Sciences (Paris)* D, 291, 187-89.

LEWIS-WILLIAMS, J.D. 1982. The economic and social context of Southern San rock art. *Current Anth.* 23, 429-49.

LEWIS-WILLIAMS, J.D. & DOWSON, T.A. 1988. The signs of all times: entoptic phenomena in Upper Palaeolithic art. *Current Anth.* 29, 201-45.

LHOTE, H. 1968. La plaquette dite de 'la Femme au Renne', de Laugerie-Basse, et son interprétation zoologique, in *Simposio de Arte Rupestre*, Barcelona 1966, 79-97.

LHOTE, H. 1968a. A propos de l'identité de la femme et du bison selon les théories récentes de l'art pariétal préhistorique, in *Ibid.*, 99-108.

LHOTE, H. 1972. Observations sur la technique et la lecture des gravures et peintures quaternaires du Sud-ouest de la France, in *Santander Symposium*, 321-30.

LINARES MALAGA, E. 1988. Arte mobiliar con tradición rupestre en el Sur del Perú. *Rock Art Research* 5 (1), 54-66.

LINDNER, K. 1950. *La Chasse Préhistorique.* Payot: Paris.

LION VALDERRABANO, R. 1971. *El Caballo en el Arte Cántabro-Aquitano.* Public. del Patronato de las cuevas prehist. de la prov. de Santander VIII: Santander.

LIPS, J.E. 1949. *The Origin of Things.* Harrap: London.

LLONGUERAS, M. 1972. Gráficos estadísticos sobre las placas de la cueva del Parpalló (Gandía, Valencia), in *Santander Symposium*, 393-405.

LOMMEL, A. 1967. *The World of the Early Hunters. Medicine-men, Shamans and Artists.* Evelyn, Adams & Mackay: London.

LORBLANCHET, M. 1974. *L'Art Préhistorique en Quercy. La Grotte des Escabasses (Thémines - Lot).* PGP: Morlaas.

LORBLANCHET, M. 1977. From naturalism to abstraction in European prehistoric rock art, in *Form in Indigenous Art* (P.J. Ucko, ed.), 44-56. Duckworth: London.

LORBLANCHET, M. 1980. Peindre sur les parois des grottes, in *Revivre la Préhistoire*, Dossier de l'Arch. 46, 33-39.

LORBLANCHET, M. 1980a. Les gravures de l'ouest australien. Leur rénovation au cours des âges. *Bull. Soc. Préhist. fr.* 77, 463-77.

LORBLANCHET, M. 1981. Les dessins noirs du Pech-Merle. in *XXIe Congrès Préhist. de France*, Montauban/Cahors 1979, vol. 1, 178-207.

LORBLANCHET, M. 1984. Les relevés d'art préhistorique, in *L'Art des Cavernes. Atlas des grottes ornées paléolithiques françaises*, 41-51. Min. de la Culture: Paris.

LORBLANCHET, M. & WELTE, A-C. 1987. Les figurations féminines stylisées du Magdalénien Supérieur du Quercy. *Bull. Soc. Etudes du Lot* 108, f.3, 3-58.

de LUMLEY, H. 1968. Le bison gravé de Ségriès, Moustiers-Ste-Marie, Bassin du Verdon (Basses-Alpes), in *Simposio de Arte Rupestre*, Barcelona 1966, 109-21.

de LUMLEY, H. 1968a. Proportions et constructions dans l'art paléolithique: le bison, in *Ibid.*, 123-45.

MACINTOSH, N.W.G. 1977. Beswick Creek cave two decades later, in *Form in Indigenous Art* (P.J. Ucko, ed.), 191-97. Duckworth: London.

MADARIAGA DE LA CAMPA, B. 1981. Historia del descubrimiento y valoración del arte rupestre español. in *Altamira Symposium*, 299-310. Min. de Cultura: Madrid.

MANIA, D. & VLČEK, E. 1987. Homo erectus from Bilzingsleben (GDR) - his culture and his environment. *Anthropologie* (Brno) 25 (1), 1-45.

MARINGER, J. & BANDI, H-G. 1953. *Art in the Ice Age.* Allen & Unwin: London.

MARSHACK, A. 1969. Polesini: a reexamination of the engraved Upper Palaeolithic mobiliary materials of Italy by a new methodology. *Rivista di Scienze Preistoriche* 24, 219-81.

MARSHACK, A. 1970. La bâton de commandement de Montgaudier (Charente): Réexamen au microscope et interprétation nouvelle. *L'Anthropologie* 74, 321-52.

MARSHACK, A. 1970a. *Notation dans les Gravures du Paléolithique Supérieur.* Mémoire 8, Inst. Préhist. Univ. Bordeaux; Delmas: Bordeaux.

MARSHACK, A. 1972. *The Roots of Civilization.* Weidenfeld & Nicolson: London.

MARSHACK, A. 1972a. Cognitive aspects of Upper Paleolithic engraving. *Current Anth.* 13, 445-77; also 15, 1974, 327-32, & 16, 1975, 297-98.

MARSHACK, A. 1975. Exploring the mind of Ice Age man. *National Geographic* 147 (1), 64-89.

MARSHACK, A. 1976. Implications of the Paleolithic symbolic evidence for the origin of language. *American Scientist* 64, 136-45.

MARSHACK, A. 1976a. Some implications of the Paleolithic symbolic evidence for the origin of language. *Current Anth.* 17, 274-81.

MARSHACK, A. 1976b. The message in the markings. *Horizon* 18, 64-73.

MARSHACK, A. 1977. The meander as a system: the analysis and recognition of iconographic units in Upper Paleolithic compositions, in *Form in Indigenous Art* (P.J. Ucko, ed.), 286-317. Duckworth: London.

MARSHACK, A. 1979. Upper Paleolithic symbol systems of the Russian Plain: cognitive and comparative analysis. *Current Anth.* 20, 271-311, 604-08.

MARSHACK, A. 1981. On Paleolithic ochre and the early uses of color and symbol. *Current Anth.* 22, 188-91.

MARSHACK, A. 1984. Concepts théoriques conduisant à de nouvelles méthodes analytiques, de nouveaux procédés de recherche et catégories de données. *L'Anthropologie* 88, 575-86.

MARSHACK, A. 1985. Theoretical concepts that lead to new analytic methods, modes of inquiry and classes of data. *Rock Art Research* 2, 95-111; and 3 (1986), 62-82 & 175-77.

MARSHACK, A. 1987. Early hominid symbol and evolution of the human capacity. Paper for *The Origin and Dispersal of Modern Humans* conference, Cambridge, March.

MARSHACK, A. 1988. La pensée symbolique et l'art, in *L'Homme de Néandertal*, Dossiers de l'Archéologie 124, 80-90.

MARTIN, Y. 1973. *L'Art Paléolithique de Gouy.* Martin: Gouy.

MARTIN, Y. 1974. Technique de moulage de gravures rupestres. *Bull. Soc. Préhist. française* 71, 146-48.

MARTIN SANTAMARIA, E. & MOURE ROMANILLO, J.A. 1981. El grabado de estilo paleolítico de Domingo García (Segovia). *Trabajos de Prehistoria* 38, 97-105, 2 pl.

MAZAK, V. 1970. On a supposed prehistoric representation of the Pleistocene scimitar cat, *Homotherium* FABRINI 1890 *(Mammalia. Machairodontidae).* Zeitschrift f. Säugetierkunde 35, 359-62.

McBURNEY, C.B.M. 1967. *The Haua Fteah (Cyrenaica) and the Stone Age of the South-east Mediterranean.* Cambridge Univ. Press.

McNICKLE, H.P. 1985. An introduction to the Spear Hill rock art complex, Northwestern Australia. *Rock Art Research* 2 (1), 48-64.

MELLARS, P.A. 1985. The ecological basis of social complexity in the Upper Palaeolithic of Southwestern France, in *Prehistoric Hunter-Gatherers: the Emergence of Social and Cultural Complexity* (T.D. Price & J.A. Brown, eds), 271-97. Academic Press: New York/London.

MESSMACHER, M. 1981. El arte paleolítico en México, in *Arte Paleolítico*, Comisión XI, Xth Congress UISPP, Mexico City, 82-110.

MESZAROS, G. & VERTES, L. 1955. A paint mine from the early Upper Palaeolithic age near Lovas (Hungary, County Veszprém). *Acta Arch. Hungarica* 5, 1-34.

MEYLAN, C. 1986. Leroi-Gourhan au bestiaire de la préhistoire. *Bull. Soc. Préhist. Ariège-Pyrénées* 41, 31-62.

MITHEN, S.J. 1988. Looking and learning: Upper Palaeolithic art and information gathering. *World Arch.* 19, 297-327.

MOLARD, Cdt. 1908. Les grottes de Sabart (Ariège): Niaux et les dessins préhistoriques. *Spelunca* VII, No. 53, 177-91.

MONS, L. 1980/1. Les baguettes demi-rondes du Paléolithique supérieur occidental: analyse et réflexions. *Antiquités Nationales* 12/13, 7-19.

MONS, L. 1986. Les statuettes animalières en grès de la grotte d'Isturitz (Pyr-Atl). Observations et hypothèses de fragmentation volontaire. *L'Anthropologie* 90, 701-11.

MONS, L. 1986/7. Les figurations de bisons dans l'art mobilier de la grotte d'Isturitz (Pyrénées-Atlantiques). *Antiquités Nationales* 18/19, 91-99.

MOORE, D.R. 1977. The hand stencil as symbol, in *Form in Indigenous Art* (P.J. Ucko, ed.), 318-24. Duckworth: London.

MORI, F. 1974. The earliest Saharan rock-engravings. *Antiquity* 48, 87-92.

MOURE ROMANILLO, J.A. 1985. Nouveautés dans l'art mobilier figuratif du Paléolithique cantabrique. *Bull. Soc. Préhist. Ariège* 40, 99-129.

MOURE ROMANILLO, J.A. 1986. New data on the chronology and context of Cantabrian Paleolithic cave art. *Current Anth.* 27, p. 65.

MOURE ROMANILLO, J.A. *et al* 1984/5. Las pinturas paleolíticas de la cueva de la Fuente del Salín (Muñorrodero, Cantabria). *Ars Praehistorica* III/IV, 13-23.

MÜLLER-BECK, H. & ALBRECHT, G. (eds). 1987. *Die Anfänge der Kunst vor 30,000 Jahren.* Theiss: Stuttgart.

MURRAY, P. & CHALOUPKA, G. 1983/4. The Dreamtime animals: extinct megafauna in Arnhem Land rock art. *Archaeology in Oceania* 18/19, 105-16.

MUZZOLINI, A. 1986. *L'Art Rupestre Préhistorique des Massifs Centraux Sahariens.* British Arch. Reports, Int. Series 318. Oxford.

NABER, F.B. *et al* 1976. *L'Art Pariétal Paléolithique en Europe Romane.* 3 vols. Bonner Hefte für Vorgeschichte Nos. 14-16.

NARR, K.J. 1960. Weibliche symbol-plastik der älteren Steinzeit. *Antaios* II (2), 132-57.

NOBBS, M. & DORN, R.I. 1988. Pleistocene age determinations for rock varnish within petroglyphs: Calibrated dates from the Olary region of South Australia. *Rock Art Research* 5 (2), in press.

NOUGIER, L-R. 1972. Nouvelles approches de l'art préhistorique animalier, in *Santander Symposium,* 263-78.

NOUGIER, L-R. 1975. L'importance du choix dans l'explication religieuse de l'art quaternaire, in *Valcamonica Symposium 1972: Les Religions de la Préhistoire,* 57-64.

NOUGIER, L-R. 1975a. Art préhistorique et topographie. *Bull. Soc. Préhist. Ariège* 30, 115-17.

NOUGIER, L-R. & ROBERT, R. 1957. *Rouffignac, ou la Guerre des Mammouths.* La Table Ronde: Paris.

NOUGIER, L-R. & ROBERT, R. 1957a. Le rhinocéros dans l'art franco-cantabrique occidental. *Bull. Soc. Préhist. Ariège* 12, 15-52, 2 pl.

NOUGIER, L-R. & ROBERT, R. 1958. Le 'lissoir aux Saïgas' de la grotte de la Vache à Alliat et l'antilope Saïga dans l'art franco-cantabrique. *Ibid.* 13, 13-28, 2 pl.

NOUGIER, L-R. & ROBERT, R. 1960. Les 'loups affrontés' de la grotte de la Vache (Ariège) et les canidés dans l'art franco-cantabrique, in *Festschrift für Lothar Zotz,* 399-420, 1 pl. Röhrscheid Verlag: Bonn.

NOUGIER, L-R. & ROBERT, R. 1965. Les Félins dans l'art quaternaire. *Bull. Soc. Préhist. Ariège* 20, 17-84; and 21, 1966, 35-46, 2 pl.

NOUGIER, L-R. & ROBERT, R. 1968. Les processions, les associations et les juxtapositions d'anthropomorphes et d'animaux dans l'art quaternaire. *Ibid.* 23, 33-98.

NOUGIER, L-R. & ROBERT, R. 1974. De l'accouplement dans l'art préhistorique. *Ibid.* 29, 15-63.

NOVEL, P. 1986. Les animaux rares dans l'art pariétal aquitain. *Bull. Soc. Préhist. Ariège-Pyrénées* 41, 63-93 (& in 42, 1987, pp. 83-118).

OBERMAIER, H. 1918. Trampas cuaternarias para espíritus malignos. *Bol. Real Soc. esp. de Hist. Nat.* 18, 162-69.

OLIVIE MARTINEZ-PEÑALVER, A. 1984/5. Representaciones de cérvidos en el arte parietal paleolítico de Cantabria. *Ars Praehistorica* 3/4, 95-110.

OLLIER DE MARICHARD, P. 1973. Un pionnier de la préhistoire ardéchoise: Jules Ollier de Marichard (1824-1901). *Etudes Préhistoriques* 4, mars, 25-29.

OMNES, J. 1982. *La Grotte Ornée de Labastide (Hautes-Pyrénées).* Omnès: Lourdes.

OTTE, M. 1974. Observations sur le débitage et le façonnage de l'ivoire dans l'Aurignacien en Belgique. in *1er Colloque Int. sur l'Industrie de l'Os dans la Préhistoire* (H. Camps-Fabrer, ed.), 93-96. Univ. de Provence.

PALES, L. 1964. Préface, in Gaussen 1964, 11-16.

PALES, L. 1969. *Les Gravures de La Marche: I, Félins et Ours.* Mémoire 7, Publications de l'Inst. de Préhistoire de Bordeaux; Delmas: Bordeaux.

PALES, L. 1970. Le 'Coco des Roseaux' ou la fin d'une erreur. *Bull. Soc. Préhist. française* 67, 85-88.

PALES, L. 1972. Les ci-devant Vénus stéatopyges aurignaciennes, in *Santander Symposium,* 217-61.

PALES, L. 1976/7. Les ovicapridés préhistoriques franco-ibériques au naturel et figurés. *Sautuola* II, 67-105, 1 pl. Publicaciones del patronato de las cuevas preh. de la Prov. de Santander, XV.

PALES, L. & de ST PEREUSE, M.T. 1965. En compagnie de l'abbé Breuil devant les bisons gravés magdaléniens de la grotte de la Marche, in *Miscelánea en Homenaje al Abate H. Breuil,* vol. II, 217-50. Barcelona.

PALES, L. & de ST PEREUSE, M.T. 1966. Un cheval-prétexte: retour au chevêtre. *Objets et Mondes* 6, 187-206.

PALES, L. & de ST PEREUSE, M.T. 1976. *Les Gravures de La Marche: II, Les Humains.* Ophrys: Paris.

PALES, L. & de ST PEREUSE, M.T. 1979. L'abri Durif à Enval (Vic-le-Comte, Puy-de-Dôme). Gravures et sculptures sur pierre. *Gallia Préhistoire* 22, 113-42.

PALES, L. & de ST PEREUSE, M.T. 1981. *Les Gravures de La Marche: III, Equidés et Bovidés.* Ophrys: Paris.

PALES, L. & de ST PEREUSE, M.T. (forthcoming) *Les Gravures de La Marche: IV, Cervidés, Eléphants et divers.*

PALES, L. et al 1976. *Les Empreintes de Pieds Humains dans les Cavernes.* Archives de l'Inst. de Paléontologie Humaine 36, Paris.

PARKINGTON, J. 1969. Symbolism in cave art. *South African Arch. Bull.* 24, 3-13.

PASSEMARD, E. & BREUIL, H. 1928. La plus grande gravure magdalénienne à contours découpés. *Rev. Arch.* 27, 4pp.

PASSEMARD, L. 1938. *Les Statuettes Féminines Paléolithiques dites Vénus stéatopyges.* Teissier: Nîmes.

PEQUART, M. & S-J. 1939/40. Fouilles archéologiques et nouvelles découvertes au Mas d'Azil. *L'Anthropologie* 49, 450-53.

PEQUART, M. & S-J. 1960/3. Grotte du Mas d'Azil (Ariège) - Une nouvelle galerie magdalénienne. *Annales de Paléontologie,* collected papers, 351pp.

PERICOT, L. 1942. *La Cueva del Parpalló (Gandía).* Consejo Sup. de Investigaciones Científicas, Inst. Diego Velázquez: Madrid.

PERICOT, L. 1962. Un curioso paralelo. *Munibe* 14, 456-58.

PERPERE, M. 1984. Les instruments de l'artiste, in *Les Premiers Artistes,* Dossier de l'Arch. 87, 41-44.

PFEIFFER, J.E. 1982. *The Creative Explosion. An inquiry into the origins of art and religion.* Harper & Row: New York.

PIETTE. E. 1887. Equidés de la période quaternaire d'après les gravures de ce temps. *Matériaux* 3e série, 4, 359-66.

PIETTE, E. 1889. L'époque de transition intermédiaire entre l'âge du Renne et l'époque de la pierre polie. *Comptes rendus du 10e Congr. Int. d'Anth. et d'Arch. Préhist.,* Paris (publ. 1891), 203-13.

PIETTE, E. 1894. Notes pour servir à l'histoire de l'art primitif. *L'Anthropologie* 5, 129-46.

PIETTE, E. 1895. La station de Brassempouy et les statuettes humaines de la période glyptique. *L'Anth.* 6, 129-51, 7 pl.

PIETTE, E. 1902. Gravure du Mas d'Azil et statuettes de Menton. *Bull. Soc. Anth. Paris* 5e série, 3, 771-79.

PIETTE, E. 1905. Les écritures de l'Age glyptique. *L'Anth.* 16, 1-11.

PIETTE, E. 1906. Le chevêtre et la semi-domestication des animaux aux temps pléistocènes. *L'Anthropologie* 17, 27-53.

PIETTE, E. 1907. *L'Art pendant l'Age du Renne.* Masson: Paris.

PITTARD, E. 1929. La première découverte d'art préhistorique (gravure et sculpture) a été faite dans la station de Veyrier (Hte-Savoie) par le Genevois François Mayor. *Revue Anth.,* 39e année, no. 7-9, 296-304.

PLENIER, A. 1971. *L'Art de la Grotte de Marsoulas.* Mémoire 1, Inst. Art Préhist. Toulouse.

POND, A.W. 1925. The oldest jewelry in the world. *Art and Archaeology* 19, 131-34, 1 pl.

POPLIN, F. 1983. Incisives de renne sciées du Magdalénien d'Europe occidentale. in *La Faune et l'Homme Préhistoriques,* Mém. 16, Soc. Préhist. française, 55-67.

POWERS, R. & STRINGER, C.B. 1975. Palaeolithic cave art fauna. *Studies in Speleology* 2 (7/8), 265-98.

PRADEL, L. 1975. Les mains incomplètes de Gargas, Tibiran et Maltravieso. *Quartär* 26, 159-66.

PRASLOV, N.D. 1985. L'art du Paléolithique Supérieur à l'est de l'Europe. *L'Anthropologie* 89, 181-92.

PRAT, F. 1986. Le cheval dans l'art paléolithique et les données de la paléontologie. *Arqueologia* (Porto) 14, dec., 27-33.

RAPHAEL, M. 1986. *L'Art Pariétal Paléolithique.* Kronos: Paris.

REGNAULT, F. 1906. Empreintes de mains humaines dans la grotte de Gargas (Hautes Pyrénées). *Bull. Soc. Anth.* 5e série, 7, 331-32.

REINACH, S. 1899. Gabriel de Mortillet. *Revue Historique* 24e année, 69, 67-95.

REINACH, S. 1903. L'art et la magie. A propos des peintures et des gravures de l'Age du Renne. *L'Anth.* 14, 257-66.

RICE, P.C. 1981. Prehistoric venuses: symbols of motherhood or womanhood? *Journal of Anth. Research* 37, 402-14.

RICE, P.C. & PATERSON, A.L. 1985. Cave art and bones: exploring the interrelationships. *American Anthropologist* 87, 94-100.

RICE, P.C. & PATERSON, A.L. 1986. Validating the cave art – archeofaunal relationship in Cantabrian Spain. *Ibid.* 88, 658-67.

RIDDELL, W.H. 1940. Dead or alive? *Antiquity* 14, 154-62.

RIPOLL PERELLO, E. 1964. Problemas cronológicos del arte paleolítico, in *Prehistoric Art of the Western Mediterranean and the Sahara* (L. Pericot & E. Ripoll, eds), 83-100. Viking Fund Publications in Anth. 39, New York.

RIPOLL PERELLO, E. 1972. *La Cueva de las Monedas en Puente Viesgo (Santander).* Monografías de Arte Rupestre, Arte Paleolítico No. 1, Barcelona.

RIPOLL PERELLO, E. 1984. Notes sur certaines représentations d'animaux dans l'art paléolithique de la péninsule ibérique, in Bandi *et al* 1984, 263-82.

RIVENQ, C. 1976. *La 'Scène de Chasse' de Ganties-Montespan.* Rivenq: Toulouse.

RIVENQ, C. 1984. Grotte de Ganties-Montespan, in *L'Art des Cavernes. Atlas des grottes ornées paléolithiques françaises*, 438-45. Min. de la Culture: Paris.

ROBERT, R. *et al* 1953. Sur l'existence possible d'une école d'art dans le Magdalénien pyrénéen. *Bull. Arch.* 187-93.

ROBINEAU, D. 1984. Sur les mammifères marins du bâton gravé préhistorique de Montgaudier. *L'Anth.* 88, 661-64.

RONEN, A. & BARTON, G.M. 1981. Rock engravings on western Mount Carmel, Israel. *Quartär* 31/32, 121-37.

ROPER, M.K. 1969. A survey of the evidence for intrahuman killing in the Pleistocene. *Current Anth.* 10, 427-59.

ROSENFELD, A. 1977. Profile figures: schematisation of the human figure in the Magdalenian culture of Europe, in *Form in Indigenous Art* (P.J. Ucko, ed.), 90-109. Duckworth: London.

ROSENFELD, A. 1984. The identification of animal representations in the art of the Laura region, North Queensland (Australia), in Bandi *et al* 1984, 399-422.

ROSENFELD, A., HORTON, D. & WINTER, J. 1981. *Early Man in North Queensland.* Terra Australis 6, Australian National University: Canberra.

ROTTLÄNDER, R. 1965. Zur Frage des Pigmentbinders der Franko-Kantabrischen Höhlenmalereien. *Fundamenta* Reihe A, Band 2, 340-44.

ROUSSEAU, M. 1967. *Les Grands Félins dans l'Art de notre Préhistoire.* Picard: Paris.

ROUSSEAU, M. 1973. Darwin et les chevaux peints paléolithiques d'Ekain. *Bull. Soc. Préhist. Ariège* 28, 49-55.

ROUSSEAU, M. 1974. Récentes déterminations et réfutations de 'Félin' dans l'art paléolithique. *Säugetierkundliche Mitteilungen* 2, 97-103.

ROUSSEAU, M. 1984. Torsions conventionnelles et flexions naturelles dans l'art animalier paléolithique et au delà, in Bandi *et al* 1984, 243-49.

ROUSSEAU, M. 1984a. Les pelages dans l'iconographie paléolithique, in *Ibid.* 161-97.

ROUSSOT, A. 1965. Les découvertes d'art pariétal en Périgord. in *Centenaire de la Préhistoire en Périgord (1864-1964)*, 99-125. Fanlac: Périgueux.

ROUSSOT, A. 1970. Flûtes et sifflets paléolithiques en Gironde. *Rev. Historique de Bordeaux et du Dépt. de la Gironde*, 5-12.

ROUSSOT, A. 1972/3. La découverte des gravures de Pair-non-Pair d'après les notes de François Daleau. *Cahiers du Vitrezais* 1, july, 5-7, & oct, 15-17; 2, april, 22-24.

ROUSSOT, A. 1981. Observations sur le coloriage de sculptures paléolithiques. *Bull. Soc. Préhist. fr.* 78, p. 200.

ROUSSOT, A. 1984. Les premiers bas-reliefs de l'humanité, in *Les Premiers Artistes*, Dossier de l'Arch. 87, 73-76.

ROUSSOT, A. 1984a. Peintures, gravures et sculptures de l'abri du Poisson aux Eyzies. Quelques nouvelles observations. *Bull. Soc. Préhist. Ariège* 39, 11-26.

ROUSSOT, A. 1984b. Approche statistique du bestiaire figuré dans l'art pariétal. *L'Anthropologie* 88, 485-98.

ROUSSOT, A. 1984c. La rondelle 'aux chamois' de Laugerie-Basse, in *Eléments de Pré et Protohistoire européenne, Hommages à J-P. Millotte*, Annales Litt. Univ. de Besançon, 219-31.

ROUZAUD, F. 1978. *La Paléospéléologie. L'Homme et le Milieu Souterrain pyrénéen au Paléolithique supérieur.* Archives d'Ecologie Préhist. 3, Toulouse.

ROZOY, J.G. 1985. Peut-on identifier les animaux de Roc-la-Tour I? *Anthropozoologica* 2, 5-7.

RUSPOLI, M. 1987. *The Cave of Lascaux: The Final Photographic Record.* Thames and Hudson: London.

RUSSELL, P.M. 1987. *Women in Upper Palaeolithic Europe.* M.A. Dissertation, University of Auckland.

SACCHI, D., ABELANET, J. & BRULE, J-L. 1987. Le rocher gravé de Fornols-Haut. *Archéologia* 225, juin, 52-57.

SAHLY, A. 1963. Nouvelles découvertes dans la grotte de Gargas. *Bull. Soc. Préhist. Ariège* 18, 65-74, 2 pl.

de SAINT MATHURIN, S. 1958. Rouffignac, ses textes, ses plans. *Bull. Soc. Préh. fr.* 55, 588-92.

de SAINT MATHURIN, S. 1971. Les biches du Chaffaud (Vienne). Vicissitudes d'une découverte. *Antiquités Nationales* 3, 22-28.

de SAINT MATHURIN, S. 1973. Bas-relief et plaquette d'homme magdalénien d'Angles-sur-l'Anglin. *Antiquités Nationales* 5, 12-19.

de SAINT MATHURIN, S. 1975. Reliefs magdaléniens d'Angles-sur-l'Anglin (Vienne). *Antiquités Nationales* 7, 24-31.

de SAINT MATHURIN, S. 1978. Les 'Vénus' pariétales et mobilières du Magdalénien d'Angles-sur-l'Anglin. *Antiquités Nationales* 10, 15-22.

de SAINT PERIER, R. 1930. La Grotte d'Isturitz, I: Le Magdalénien de la Salle de St-Martin. *Archives de l'Inst. Pal. Humaine* 7, Paris.

de SAINT PERIER, R. 1936. La Grotte d'Isturitz, II: Le Magdalénien de la Grande Salle. *Archives de l'Inst. Pal. Humaine* 17, Paris.

de SAINT PERIER, R-S. 1965. Inventaire de l'art mobilier paléolithique du Périgord. in *Centenaire de la Préhistoire en Périgord (1864-1964)*, 139-59. Fanlac: Périgueux.

de SAUTUOLA, M.S. 1880. *Breves apuntes sobre algunos objetos prehistóricos de la provincia de Santander.* Santander, 27pp, 4 pl.

SAUVET, G. 1983. Les représentations d'équidés paléolithiques de la grotte de la Griega (Pedraza, Segovia). A propos d'une nouvelle découverte. *Ars Praehistorica* 2, 49-59.

SAUVET, G. & S. 1979. Fonction sémiologique de l'art pariétal animalier franco-cantabrique. *Bull. Soc. Préhist. fr.* 76, 340-54.

SAUVET, G. & S., & WLODARCZYK, A. 1977. Essai de sémiologie préhistorique (pour une théorie des premiers signes graphiques de l'homme). *Ibid.* 74, 545-58.

SCHMID, E. 1984. Some anatomical observations on Palaeolithic depictions of horses, in Bandi *et al* 1984, 155-60.

SHIMKIN, E.M. 1978. The Upper Palaeolithic in North-Central Eurasia: evidence and problems, in *Views of the Past* (L.G. Freeman, ed.), 193-315. Mouton: The Hague.

SIEVEKING, A. 1971. Palaeolithic decorated bone discs. *British Museum Quarterly* 35, 206-29, 11 pl.

SIEVEKING, A. 1978. La significación de las distribuciónes en el arte paleolítico. *Trabajos de Prehistoria* 35, 61-80.

SIEVEKING, A. 1979. *The Cave Artists.* Thames & Hudson: London.

SIEVEKING, A. 1979a. Style and regional grouping in Magdalenian cave art. *Bull. Inst. of Arch.* 16, 95-109.

SIEVEKING, A. 1983. Decorated scapulae from the Upper Palaeolithic of France and Cantabrian Spain, in *Homenaje al Prof. M. Almagro Basch*, vol. 1, 313-17. Min. de Cultura: Madrid.

SIEVEKING, A. 1987. *Engraved Magdalenian Plaquettes.* British Arch. Reports Int. series No. 369: Oxford.

SIEVEKING, A. & G. 1962. *The Caves of France and Northern Spain.* Vista Books: London.

SIEVEKING, G. 1972. Art mobilier in Britain, in *Santander Symposium,* 385-88.

SIMEK, J. 1986. A Paleolithic sculpture from the abri Labattut in the American Museum of Natural History collection. *Current Anth.* 27, 402-07.

SIMONNET, G. 1952. Une belle parure magdalénienne, in *12e Congrès Préhist. de France,* Paris 1950, 564-68.

SIMONNET, G., L. & R. 1984. Quelques beaux objets d'art venant de nos recherches dans la grotte ornée de Labastide (H-P). Approche naturaliste. *Bull. Soc. Méridionale de Spéléol. et Préhist.* 24, 25-36.

SIMONNET, R. 1980. Emergence de la préhistoire en pays ariégeois. Aperçu critique d'un siècle de recherches. *Bull. Soc. ariégeoise Sciences, Lettres et Arts* 35, 5-88.

SMITH, N.W. (in press) *A Psychology of Ice Age Art: An Analysis.* Stanford Univ. Press.

SOFFER, O. 1985. *The Upper Paleolithic of the Central Russian Plain.* Academic Press: Orlando.

SOFFER, O. 1985a. Patterns of intensification as seen from the Upper Paleolithic of the Central Russian Plain, in *Prehistoric Hunter-Gatherers. The Emergence of Cultural Complexity* (T.D. Price & J.A. Brown, eds), 235-70. Academic Press: New York/London.

SOHN POW-KEY 1974. Palaeolithic culture of Korea. *Korea Journal,* april, 4-11.

SOHN POW-KEY 1981. Inception of art mobilier in the Middle Palaeolithic period at Chommal Cave, Korea. in *Resumenes de Comunicaciones, Paleolítico Medio,* Xth Congress UISPP, Mexico City, 31-32.

de SONNEVILLE-BORDES, D. 1986. Le bestiaire paléolithique en Périgord. Chronologie et signification. *L'Anthropologie* 90, 613-56.

de SONNEVILLE-BORDES, D. & LAURENT, P. 1983. Le phoque à la fin des temps glaciaires, in *La Faune et l'Homme Prèhistoriques, Mém. Soc. Préhist. fr.* 16, 69-80.

de SONNEVILLE-BORDES, D. & LAURENT, P. 1986/7. Figurations de chevaux à l'abri Morin. Observations complémentaires. *Antiquités Nationales* 18/19, 69-74.

SOURIAU, E. 1971. Art préhistorique et esthétique du mouvement, in *Mélanges A. Varagnac,* 697-705. Sevpen: Paris.

SPECK, F.G. & SCHAEFFER, C.E. 1950. The deer and rabbit hunting drive in Virginia. *Southern Indian Studies* 2, 3-20.

STEVENS, A. 1975. Animals in Palaeolithic cave art: Leroi-Gourhan's hypothesis. *Antiquity* 49, 54-57.

STEVENS, A. 1975a. Association of animals in Palaeolithic cave art: The second hypothesis. *Science and Archaeology* 16, 3-9.

STOLIAR, A.D. 1977/8. On the genesis of depictive activity and its role in the formation of consciousness (Toward a formulation of the problem). *Soviet Anth. & Arch.* 16, 3-42; & 17 (1978/9), 3-33.

STOLIAR, A.D. 1977/8a. On the sociohistorical decoding of Upper Paleolithic female signs. *Ibid.* 16, 36-77.

STOLIAR, A.D. 1981. On the archeological aspect of the problem of the genesis of animalistic art in the Eurasian Paleolithic. *Ibid.* 20, 72-108.

STOLIAR, A.D. 1985. *Proisxozdenie izobrazitel'nogo iskusstva* (The origin of figurative art). Iskusstvo: Moscow.

STRAUS, L.G. 1982. Observations on Upper Paleolithic art: old problems and new directions. *Zephyrus* 34/5, 71-80.

STRAUS, L.G. 1987. The Paleolithic cave-art of Vasco-Cantabrian Spain. *Oxford Journal of Arch.* 6, 149-63.

TABORIN, Y. 1977. Quelques objets de parure. Etude technologique: les percements des incisives de bovinés et des canines de renards. *Méthodologie Appliquée à l'Industrie de l'Os Préhistorique* (H. Camps-Fabrer, ed), 303-10. C.N.R.S.: Paris.

TABORIN, Y. 1982. La parure des morts. *La Mort dans la Préhistoire*, Dossier de l'Archéologie 66, 42-51.

TABORIN, Y. 1985. Les origines des coquillages paléolithiques en France. in *La Signification culturelle des Industries Lithiques* (M. Otte, ed), 278-301. British Arch. Reports Int. series No. 239: Oxford.

TARASSOV, L.M. 1971. La double statuette paléolithique de Gagarino. *Quartär* 22, 157-63, 1 pl.

THEVENIN, A. 1983. Les galets gravés et peints de l'abri de Rochedane (Doubs) et le problème de l'art azilien. *Gallia Préhistoire* 26, 139-88.

TOSELLO, G. 1983. *Inventaire des Sites Paléolithiques d'Art Pariétal et Mobilier Figuratif.* Maîtrise de Préhistoire, 2 vols. Univ. de Paris I, Panthéon-Sorbonne.

TWIESSELMANN, F. 1951. *Les Représentations de l'Homme et des Animaux Quaternaires découvertes en Belgique.* Mémoire 113, Institut Royal des Sciences Naturelles de Belgique: Brussels.

TYLDESLEY, J.A. & BAHN, P.G. 1983. Use of plants in the European Palaeolithic: a review of the evidence. *Quaternary Science Reviews* 2, 53-81.

UCKO, P.J. 1977. Opening remarks, in *Form in Indigenous Art*, (P.J. Ucko, ed.), 7-10. Duckworth: London.

UCKO, P.J. & LAYTON, R. 1984. La subjectivité et le recensement de l'art paléolithique. Typescript for Colloque Int. d'Art Pariétal Paléolithique, Périgueux. 39pp.

UCKO, P.J. & ROSENFELD, A. 1967. *Palaeolithic Cave Art.* World Univ. Library: London.

UCKO, P.J. 1987. Débuts illusoires dans l'étude de la tradition artistique. *Bull. Soc. Préhist. Ariège-Pyrénées* 42, 15-81.

UCKO, P.J. & ROSENFELD, A. 1972. Anthropomorphic representations in Palaeolithic art, in *Santander Symposium* 149-215.

UTRILLA, P. 1979. Acerca de la posición estratigráfica de los cervidos y otros animales de trazo múltiple en el Paleolítico Superior español. *Caesaraugusta* 49/50, 65-72.

VALLOIS, H.V. 1961. The social life of early man: the evidence of skeletons, in *Social Life of Early Man* (S.L. Washburn, ed.), 214-35. Viking Fund Publics. No. 31, New York.

VALOCH, K. 1970. Oeuvres d'art et objets en os du Magdalénien de Moravie (Tchécoslovaquie). *Bull. Soc. Préhist. Ariège* 25, 79-93.

VAN NOTEN, F. 1977. Excavations at Matupi Cave. *Antiquity* 51, 35-40.

VANDIVER, P. (in press) *Paleolithic Pigment Processing: a soft stone technology.* Univ. of Chicago Press.

VASIL'EV, S.A. 1985. Une statuette d'argile paléolithique de Sibérie du sud. *L'Anthropologie* 89, 193-96.

VAYSON de PRADENNE, A. 1934. Les figurations d'oiseaux dans l'art quaternaire. *I.P.E.K.* 3-17.

VERTUT, J. 1980. Contribution des techniques photographiques à l'étude et à la conservation de l'art préhistorique, in *Altamira Symposium*, 661-75. Min. de Cultura: Madrid.

VEZIAN, J. 1956. Les utilisations de contours de la roche dans la grotte du Portel. *Bull. Soc. Préhist. Ariège* 11, 79-87, 4 pl.

VIALOU, D. 1976. *Guide des Grottes Ornées Paléolithiques Ouvertes au Public.* Masson: Paris.

VIALOU, D. 1981. L'art préhistorique: questions d'interprétations. *Monuments Historiques* 118, Nov/Déc., 75-83.

VIALOU, D. 1982. Une lecture scientifique de l'art préhistorique. *La Recherche* 13, 1484-87.

VIALOU, D. 1982a. Niaux, une construction symbolique magdalénienne exemplaire. *Ars Praehistorica* 1, 19-45.

VIALOU, D. 1984. Les cervidés de Lascaux, in *Bandi et al* 1984, 199-216.

VIALOU, D. 1986. *L'Art des Grottes en Ariège Magdalénienne.* XXIIe Supplément à Gallia Préhistoire, 28 pl.

VIALOU, D. 1987. D'un tectiforme à l'autre, in *Sarlat en Périgord*, Actes du 39e Congr. d'Etudes Régionales, Suppl. to *Bull. Soc. Hist. et Arch. du Périgord* 114, 307-17.

WAKANKAR, V.S. 1984. Bhimbetka and dating of Indian rock paintings, in *Rock Art of India* (K.K. Chakravarti, ed.), 44-56. Arnold-Heinemann: New Delhi.

WAKANKAR, V.S. 1985. Bhimbetka: the stone tool industries and rock paintings, in *Recent Advances in Indo-Pacific Prehistory* (V.N. Misra & P. Bellwood, eds.), 175-76. Oxford & IBH Publishing Co: New Delhi.

WALKER, N.J. 1987. The dating of Zimbabwean rock art. *Rock Art Research* 4, 137-49.

WALSH, G.L. 1979. Mutilated hands or signal stencils. *Australian Archaeology* 9, 33-41.

WEISSEN-SZUMLANSKA, M. 1951. A propos des gravures et peintures rupestres. *Bull. Soc. Préhist. fr.* 48, 457-59.

WELTE, A-C. 1975/6. L'Affrontement dans l'art préhistorique. *Travaux de l'Inst. d'Art Préhist. Toulouse* 17, 207-305 & 18, 187-330.

WENDT, W.E. 1974. 'Art mobilier' aus der Apollo 11-Grotte in Südwest-Afrika. *Acta Praehistorica et Archaeologica* 5, 1-42.

WENDT, W.E. 1976. 'Art mobilier' from the Apollo 11 cave, South West Africa: Africa's oldest dated works of art. *South African Archaeological Bulletin* 31, 5-11.

WHITE, R. 1986. *Dark Caves, Bright Visions, Life in Ice Age Europe.* Exhibition Catalogue, American Museum of Natural History: New York.

WHITNEY, D.S. & DORN, R.I. 1987. Rock art chronology in eastern California. *World Arch.* 19 (2), 150-64.

WILDGOOSE, M., HADINGHAM, E. & HOOPER, A. 1982. The prehistoric hand pictures at Gargas: attempts at simulation. *Medical History* 26, 205-7.

WORSAAE, J.J.A. 1869. Communication...relative à l'authenticité des os de renne présentant des dessins et trouvés en France dans les cavernes du Périgord. *Congrès Int. d'Anth. et d'Arch. Préhistoriques*, 4e session, Copenhagen, 127-34. (Publ. 1875).

WRESCHNER, E. 1975. Ochre in prehistoric contexts. Remarks on its implications to the understanding of human behavior. *Mi'Tekufat Ha'Even* 13, 5-11.

WRESCHNER, E. 1980. Red ochre and human evolution: a case for discussion. *Current Anth.* 21, 631-44.

WRIGHT, B. 1985. The significance of hand motif variations in the stencilled art of the Australian Aborigines. *Rock Art Research* 2, 3-19.

WRIGHT, R.V.S. (ed.) 1971. *Archaeology of the Gallus Site, Koonalda Cave.* Australian Institute of Aboriginal Studies: Canberra.

YATES, F.A. 1966. *The Art of Memory.* Univ. of Chicago Press.

YOU YUZHU 1984. Preliminary study of a Palaeolithic bone engraving. *Kexue Tongbao* 29 (1), 80-82.

ZERVOS, C. 1959. *L'Art de l'Epoque du Renne en France.* Cahiers d'Art: Paris.

ZEUNER, F.E. 1954. The colour of the wild cattle of Lascaux. *Man* 53, 68-69.

ZHUROV, R.I. 1977/8. On one hypothesis relating to the origin of art. *Soviet Anth. & Arch.* 16, 43-63.

ZHUROV, R.I. 1981/2. On the question of the origin of art (an answer to opponents). *Soviet Anth. & Arch.* 20 (3), 59-82.

ZÜCHNER, C. 1975. Der Bison in der eiszeitlichen Kunst Westeuropas. *Madrider Mitteilungen* 16, 9-24.

Illustrations

Index

Numbers in italics refer to pages on which illustrations appear